The contents of this publication are solely those of the authors and contributors, and not of the publisher, editor(s), or employees of Silver Rock Publishing. Silver Rock Publishing, its parent company, and employees disclaim all responsibility for any injury or adverse effects of any kind to persons or property resulting from any ideas, methods, instructions, or products referred to in this publication. We present this book solely as an historical document and do not condone or endorse violence of any kind. Quite to the contrary, Silver Rock Publishing urges anyone considering acts of violence as a solution to seek professional assistance immediately.

U.S. ARMY SPECIAL FORCES GUIDE TO UNCONVENTIONAL WARFARE

Devices and Techniques for Incendiaries

Department of the Army

Published by Silver Rock Publishing

ISBN (Paperback): 978-1-62654-270-9
ISBN (Hardcover): 978-1-62654-271-6
ISBN (Spiral): 978-1-62654-272-3

INTRODUCTION

This manual is one of a series now being published, primarily for U.S. Army Special Forces, that deals with subjects pertaining to destructive techniques and their applications to targets in guerrilla and unconventional warfare. The series consists of both classified and unclassified manuals of three types:

 a. Unconventional Warfare Reference Manuals consist of detailed, illustrated abstracts of the technical literature. They are designed to provide sources of information and ideas and to minimize duplication of effort.

 b. Unconventional Warfare Devices and Technique Manuals cover incendiaries, explosives, weapons, and harmful additives. They present principles of construction and methods of use of devices and techniques that are proven reliable and effective.

 c. Unconventional Warfare Target Manuals identify critical components of selected targets and describe methods for destruction of the target using applicable devices and techniques.

This manual on incendiaries is written to serve the U.S. Army Special Forces in the field. It covers all aspects of incendiary systems including the incendiary devices, means for igniting them, techniques for their use, methods of improvising them, and sources of material supply. Some of the devices can be improvised from locally available materials. Detailed instructions are given for preparation steps. Others can be improvised if more sophisticated materials are obtainable. Still others require fabrication or formulation in a laboratory or industrial plant.

All of the devices and techniques herein reported are known to work. It is a special feature of this manual to present only those items that produce useful results as verified by independent test. Before an item becomes eligible for inclusion in this manual, it passes engineering tests designed to evaluate effectiveness, reliability, and safety. Not only does this program provide performance data, it also eliminates items that prove to be ineffective, unreliable, or unsafe, even though they may exist in print elsewhere. Although laboratory and final testing are adequate, user familiarization with construction, operation, and performance of each item or technique is recommended before tactical use. Instructions and formulations must be followed precisely to assure proper functioning of the incendiaries.

The material in this manual is grouped into six chapters.

Each chapter is subdivided into paragraphs having 4-digit numbers, the first representing section numbers, and the last two paragraph numbers.

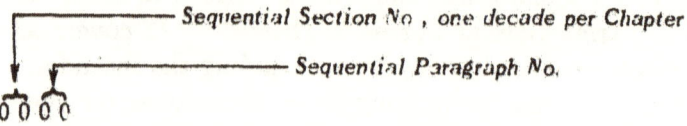

This numbering system was selected to make location of material convenient for the reader, once he has become familiar with the arrangement. Section numbers are uniquely assigned to one subject and repeated for the same subject in other manuals of this series. Also for convenience, each paragraph (except in *Introduction*) is subdivided uniformly into four subparagraphs:

Description, Material and Equipment, Preparation, and Application.

It is anticipated that this manual will be revised or changed from time to time. In this way it will be possible to update present material and add new devices and techniques as they become available. Comments are invited and the submission of new information is encouraged. Address information to Commanding Officer, U.S. Army Frankford Arsenal, SMUFA-U3200, Philadelphia, Pa., 19137.

TM 31-201-1

Technical Manual } HEADQUARTERS
No. 31-201-1 } DEPARTMENT OF THE ARMY
Washington, D.C., *20 May 1966*

Unconventional Warfare Devices and Techniques
INCENDIARIES

	Paragraph	Page
Chapter 1. INTRODUCTION		
Incendiary systems	0001	3
Definitions	0002	4
Tools and techniques	0003	5
2. INITIATORS		
Fuse cord	0101	8
Improvised string fuse	0102	14
Concentrated sulfuric acid	0103	16
Water	0104	21
3. IGNITERS		
Sugar-chlorate	0201	23
Fire fudge	0202	25
Sugar—sodium peroxide	0203	28
Aluminum powder—sodium peroxide	0204	30
Match head	0205	32
Potassium permanganate—glycerin	0206	34
Powdered aluminum—sulfur pellets	0207	36
Silver nitrate—magnesium powder	0208	39
White phosphorous	0209	41
Magnesium powder—barium peroxide	0210	45
Subigniter for thermite	0211	47

TAGO 7189-C—June

	Paragraph	Page
CHAPTER 4. INCENDIARY MATERIALS		
Napalm	0301	50
Gelled gasoline (exotic thickeners)	0302	53
Gelled gasoline (improvised thickeners)	0303	57
Paraffin-sawdust	0304	76
Fire bottle (impact ignition)	0305	78
Fire bottle (delay ignition)	0306	82
Thermite	0307	85
Flammable liquids	0308	90
Incendiary brick	0309	91
5. DELAY MECHANISMS		
Cigarette	0401	94
Gelatin capsule	0402	98
Rubber diaphragm	0403	101
Paper diaphragm (sulfuric acid)	0404	105
Paper diaphragm (glycerin)	0405	107
Candle	0406	109
Overflow	0407	111
Tipping delay—filled tube	0408	113
Tipping delay—corrosive or dissolving action	0409	117
Balancing stick	0410	121
Stretched rubber band	0411	126
Alarm clock	0412	128
6. SPONTANEOUS COMBUSTION		
Spontaneous combustion	0501	131
INDEX		143

CHAPTER 1
INTRODUCTION

0001. INCENDIARY SYSTEMS

  a. This manual covers all aspects of incendiary systems. It describes useful initiators, igniters, incendiary materials, delay mechanisms, and spontaneous combustion devices designed for direct use in sabotage and unconventional warfare.

  b. Incendiaries are primarily used in sabotage to set fire to wooden structures and other combustible targets. Certain incendiaries, such as thermite, can be used for melting, cutting, or welding metals.

  c. The most basic incendiary system consists of putting a lighted match to an easily combustible material. However, a simple match is not always effective. There are many important combustible targets that require far more heat for reliable ignition than is available from a match. There are also instances where delayed ignition is essential for sabotage success. This manual contains formulations and devices to satisfy the requirements for high ignition heat and predetermined ignition delay times for use in sabotage and other harassment actions.

  d. Every incendiary system consists of a group of elements starting with an initiator and ending with the main incendiary material. If the initiator does not produce enough heat for reliable ignition of the combustible target, an intermediate or booster incendiary is required. More than one booster is necessary for some targets. The initiator (ch 2) can consist of a simple match, a match and a fuse cord, an acid, or water. The intermediate heat sources are generally called igniters (ch 3). Igniters produce sufficient heat to set the principal incendiary charge (ch 4) aflame. Delay mechanisms (ch 5) are frequently used to prevent detection of the saboteur by postponement of the fire for some limited, predetermined time after placement and actuation of the device.

e. Spontaneous combuston is a good sabotage tool. Favorable conditions can be established for the deliberate employment of spontaneous combustion (ch 6), that is, setting combustible material aflame without application of direct flame or spark.

f. All of the devices and techniques described herein have been thoroughly checked by independent test to make certain that they work as intended. Detailed instructions are given for the necessary ingredients and their preparation. It is important that these instructions be followed carefully to be sure that the devices will operate properly. In addition, close attention to the instructions will assure safety.

0002. DEFINITIONS

Common terms used in connection with incendiary systems are defined below. Note that the definitions are worded so as to cover only incendiaries. Some of the terms have additional meanings in the related field of explosives.

a. Delay Mechanism. Chemical, electrical, or mechanical elements that provide a time delay. Elements may be used singly or in combination. They provide a predetermined, limited time interval before an incendiary starts to burn.

   b. Fuse. A flexible fabric tube containing powder that is used to start fires at some remote location. The powder in the fuse burns and provides a time delay.

   c. Igniter. An intermediate charge between an initiator and an incendiary material. It is set aflame by the initiator and produces sufficient heat at high temperature to ignite the main incendiary. Igniters are fast burning and relatively short lived.

   d. Incendiary Material. A material that burns with a hot flame for long periods. Its purpose is to set fire to wooden structures and other combustible targets.

   e. Incendiary System. A group of elements that are assembled to start fires. The system consists of initiator, delay mechanism (if needed), igniter, and incendiary material.

   f. Initiator. The source that provides the first fire in an incendiary system. A match is an initiator. The initiator is so sensitive that it can be set off with little energy.

   g. Spontaneous Combustion. The outbreak of fire in combustible material that occurs without an application of direct spark or flame. The fire is the result of heat produced by the chemical action of certain oils.

   h. Thermite. An incendiary mixture of iron oxide flakes and aluminum powder that reacts chemically when initiated to form molten iron. Thermite can be used to burn holes in steel or to weld steel parts together.

0003. TOOLS AND TECHNIQUES

   a. The equipment needed for the manufacture of incendiaries consists of simple items. They are all readily available. Required are bottles, jars, pots, and spoons. There should be no difficulty in obtaining any of them. All of the necessary equipment is described in each paragraph dealing with a particular incendiary component.

b. It is important that the operator follow the directions given in this manual *exactly* as written. They have been worked out carefully to give the desired results with the minimum chance of mishap. Don't experiment with different procedures or quantities.

c. By its very nature, the manufacture of incendiaries is dangerous. It is the function of incendiaries to burn with an intense flame under the right conditions. Care must be taken that no fires result during the making or placing of the devices. There are also other dangers in addition to the fire hazard. The chemicals used as ingredients may burn the skin, give off poisonous fumes, or be easily flammable. They must not be eaten.

d. When handled with care and proper precautions, incendiaries are fairly safe to make and use. Detailed precautions and instructions are given in each paragraph where they apply. General safety precautions follow:

Preventing a Fire Hazard

1. Fire prevention is much more important than fire fighting. Prevent fires from starting.

2. Keep flammable liquids away from open flames.

3. Good housekeeping is the fire prevention. Keep work areas neat and orderly. Clean away all equipment and material not needed at the moment. Clean up spills as soon as possible.

4. Store incendiaries in closed containers away from heat. Do not store material any longer than necessary.

5. In the event of fire, remove the incendiaries from the danger area if this can be done quickly and safely. Use large quantities of water to fight fires.

6. Horse play is dangerous and absolutely intolerable.

Avoiding Chemical Hazards

1. Wear rubber gloves, apron, and glasses when handling concentrated chemicals if at all possible.

2. Avoid inhaling fumes. Perform reactions in a well ventilated area or out of doors because the boiling is often violent and large amounts of fumes are given off that are poisonous if breathed too much.

3. Avoid acid contact with the skin. If chemicals are spilled on a person, wash immediately in running water for several minutes. If they splash in the eyes, wash the open eye in running water for at least 15 minutes.

4. Clean up any acid that is spilled on floor or bench by flushing with large amounts of water. Acid spilled on wood can cause a fire.

5. Always pour concentrated acids into water. Never pour water into concentrated acids because a violent reaction will occur.

CHAPTER 2
INITIATORS

0101. FUSE CORD

a. *Description.*

(1) This item consists of a continuous train of explosive or fastburning material enclosed in a flexible waterproof cord or cable. It is used for setting off an explosive or a combustible mixture of powders by action of the fuse flame on the material to be ignited. Fuse cord can be initiated by a match flame, using a specific procedure, or with a standard U.S. Army fuse lighter. Fuse cord burns at a uniform rate allowing the user to be away from the immediate scene when the incendiary actually functions.

(2) Fuse cord does not directly ignite any incendiaries listed in chapter 4 but is a primary initiator for all igniters listed in chapter 3 except: Potassium Permanganate—Glycerin (0206), Powdered Aluminum—Sulfur Pellets (0207), White Phosphorus (0209), and Sub-igniter For Thermite (0211).

b. *Material and Equipment.* Two Standard U.S. Army fuse cords are available:

(1) *Blasting time fuse.*

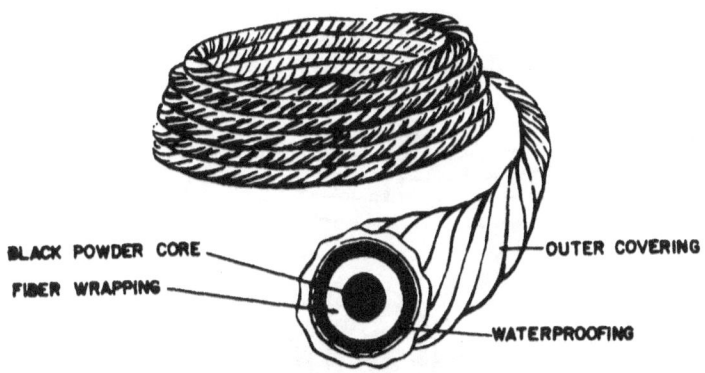

This consists of black powder tightly wrapped with several layers of fabric and waterproofing materials. It might be any color, orange being the most common. The diameter of this fuse cord is 0.2 inch (a little larger than 3/16 inch). This fuse burns inside the wrapping at a rate of approximately 40 seconds per foot. It must be tested before use to verify the burning rate.

(2) *Safety fuse M700.*

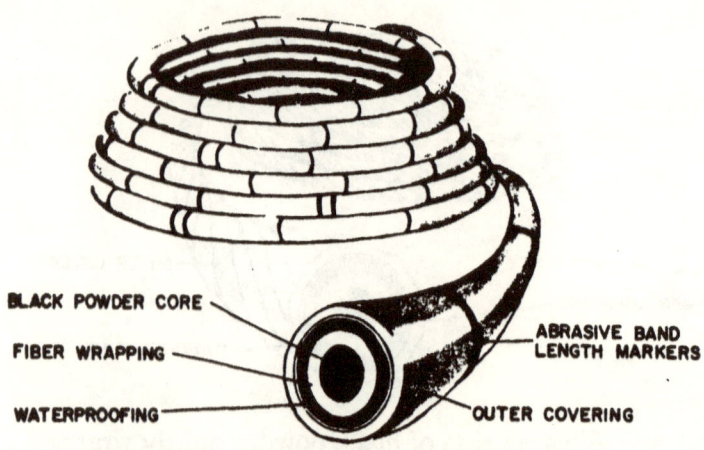

This fuse is similar to Blasting Time Fuse and may be used interchangeably with it. The fuse is a dark green cord 0.2 inch in diameter with a plastic cover, either smooth or with single painted abrasive bands around the outside at 1 foot or 18 inch intervals and double painted abrasive bands at 5 foot or 90 inch intervals depending on the time of manufacture. These bands are provided to make measuring easy. They are abrasive so that they can be felt in the dark. The fuse burns inside the wrapping at a rate of approximately 40 seconds per foot. It must be tested before use to verify the burning rate.

Note. A commercial item can be substituted for either of the above U.S. Army issue items. The commercial fuse is 0.1 inch (about 3/32 inch) in diameter and is coated only with waterproofing lacquer. This fuse can be easily ignited by holding the free end in a match flame because the outside covering if flammable.

c. Preparation. None.

d. Application.

(1) *General.*

   (a) Cut and discard a 6-inch length from the free end of the fuse roll. Do this to be sure that there is no chance of misfire from a damp powder train because of absorption of moisture from the open air. Then cut off a measured length of fuze to check the burning rate. Check the burning rate before actual use.

   (b) Cut the fuse long enough to allow a reasonable time delay in initiation of the incendiary system. The cut should be made squarely across the fuse.

   (c) Prepare the fuse for ignition by splitting the fuse at one end to a depth of about one inch. Place the head of an unlighted match in the powder train.

   (d) Insert the other end of the fuse into a quantity of an igniter mixture so that the fuse end terminates near the center of the mixture. Be sure the fuse cord is anchored in the igniter mixture and cannot pull away. In the case of a solid igniter material such as Fire Fudge (0202), the fuse is split to about one-half inch at the end opposite the end containing the match in the powder train. This split fuse end is wedged over a sharp edge of the solid igniter material. Be sure the black powder in the fuse firmly contacts the solid igniter. If necessary, the fuse cord can be held firmly to the solid igniter with light tape such as transparent adhesive tape.

(e) The fuse is initiated by lighting the match head inserted in the split end of the fuse with a burning match as shown below.

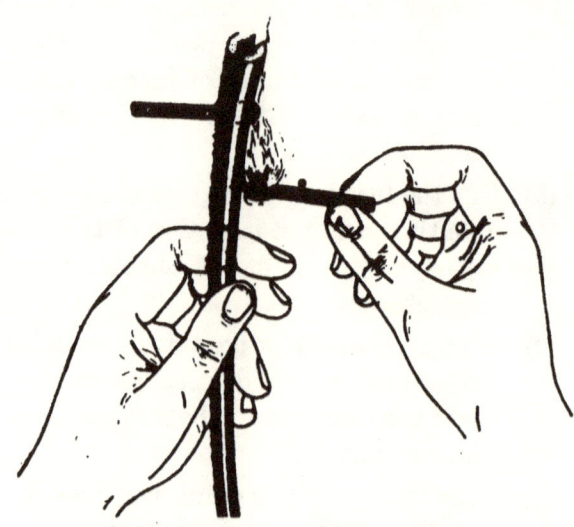

(f) Two standard fuse lighters, the M2 and M60, are available in demolition kits for positive lighting of Blasting Time Fuse and Safety Fuse M700 under all weather conditions—even under water if it is properly waterproofed. The devices are manually operated. A pull on the striker retaining pin causes the striker to hit the percussion primer, thus igniting the fuse. *These devices are not recommended where silence is required because a report is heard when the primer is fired.*

(2) *M2 fuse lighter.*

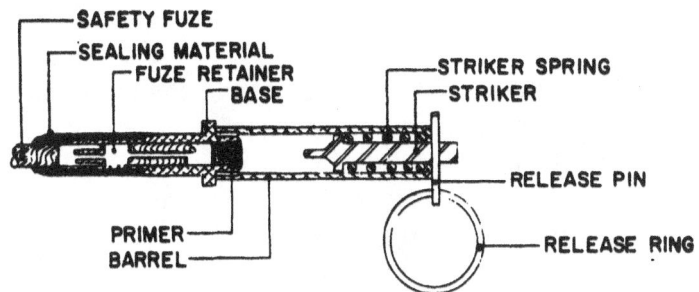

The attachment and operation of the M2 Fuse Lighter are as follows:
 (a) Slide the pronged fuse retainer over the end of the fuse and firmly seat it.
 (b) Waterproof the joint between the fuse and the lighter, if necessary, by applying a sealing compound (putty or mastic).
 (c) In firing, hold the barrel in one hand and pull on the release pin with the other hand.

(3) *M60 fuse lighter.*

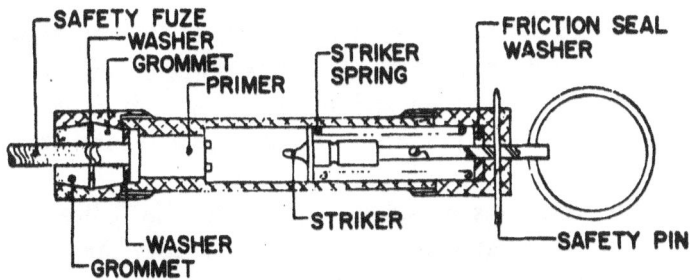

The attachment and operation of the M60 Fuse Lighter are as follows:

(a) Unscrew the fuse holder cap two or three turns.
(b) Press the shipping plug into the lighter to release the split grommet, and rotate the plug as it is removed.
(c) Insert end of fuse in place of the plug until it rests against the primer.
(d) Tighten the fuse holder cap sufficiently to hold the fuse tightly in place and thus waterproof the joint.
(e) To fire, remove the safety pin, hold the barrel in one hand, and pull on the pull ring with the other hand.

102. IMPROVISED STRING FUSE

a. *Description.*

(1) This item consists of string, twine, or shoelaces that have been treated with either a mixture of potassium nitrate and granulated sugar or potassium chlorate and granulated sugar.
(2) Improvised string fuse does not directly ignite any incendiaries listed in chapter 4 but is a primary initiator for all igniters listed in chapter 3 except: Potassium Permanganate —Glycerin (0206), Powdered Aluminum—

Sulfur Pellets (0207), White Phosphorus (0209), and Subigniter For Thermite (0211).
 (3) Depending upon the length of the fuse, the user can be away from the immediate scene when an incendiary system is initiated.
b. *Material and Equipment.*
 String, twine or shoelaces made of cotton or linen.
 Potassium nitrate or potassium chlorate.
 Granulated sugar.
 Small cooking pot.
 Spoon.
 Heat source such as stove or hot plate.
 Soap.
c. *Preparation.*
 (1) Wash string or shoelaces in hot soapy water; rinse in fresh water.
 (2) Dissolve one part potassium nitrate or potassium chlorate and one part granulated sugar in two parts hot water.
 (3) Soak string or shoelaces in the hot solution for at least five minutes.
 (4) Remove the string from hot solution and twist or braid three strands of string together.
 (5) Hang the fuse up to dry.
 (6) Check actual burning rate of the fuse by measuring the time it takes for a known length to burn.
d. *Application.*
 (1) This fuse does not have a waterproof coating and it must be tested by burning a measured length before actual use.
 (2) Cut the fuse long enough to allow a reasonable time delay in initiation of the incendiary system.

(3) Insert one end of the fuse in a quantity of an igniter mixture so that the fuse end terminates near the center of the mixture. Be sure the fuse cord is anchored in the igniter mixture and cannot pull away. In the case of a solid igniter material such as Fire Fudge (0202), the improvised string fuse is securely wrapped around a piece of solid igniter material.

(4) The fuse is initiated by lighting the free end of the fuse with a match.

(5) This fuse does not burn when it is wet. Its use is not recommended where there is the possibility of the fuse getting wet.

0103. CONCENTRATED SULFURIC ACID (OIL OF VITRIOL)

a. *Description.*

(1) This material is a heavy, corrosive, oily, and colorless liquid. Storage is recommended in a glass container with a glass lid or stopper. Commercially available sulfuric acid is approximately 93 percent concentration with a specific gravity of 1.835. This is commonly referred to as concentrated sulfuric acid.

(2) Concentrated sulfuric acid chars wood, cotton, and vegetable fibers, usually without causing fire. The addition of water to concentrated sulfuric acid develops much heat which may be sufficient to cause a fire or an explosion. This depends upon the quantity of acid, quantity of water, and rate of addition of water.

Caution: **Always add concentrated sulfuric acid to water. Never add water to a concentrated acid.**

(3) Certain igniter materials can be reliably brought to flaming by the addition of concentrated sulfuric acid. This is brought about by the chemical reaction between the sulfuric acid and the igniter materials. The following igniters are initiated by concentrated sulfuric acid: Sugar-Chlorate (0201), Fire Fudge (0202), Sugar—Sodium Peroxide (0203), Aluminum Powder—Sodium Peroxide (0204), Match Head (0205), and Silver Nitrate—Magnesium Powder (0208).

(4) The most important use for concentrated sulfuric acid as an initiator is in conjunction with delay mechanisms. The acid is held away from the igniter for a period of time by making use of the corrosive action of the acid to work its way through a barrier. If the delay mechanism is placed in a cold environment, the concentrated acid will remain fluid at extremely low temperatures. The following delay mechanisms are recommended for use with concentrated sulfuric acid: Gelatin Capsule (0402), Rubber Diaphragm (0403), Paper Diaphragm (0404), Tipping Delay—Filled tube (0408), Tipping Delay—Balancing Stick (0410), and Stretched Rubber Band (0411).

b. *Material and Equipment.* Concentrated sulfuric acid.

c. *Preparation.* None—If only battery-grade sulfuric acid is available (specific gravity 1.200), it must be concentrated before use to a specific gravity of 1.835. This is done by heating it in an enameled, heat-resistant glass or porcelain pot until dense, white fumes appear. Heat only in a well ventilated area. When dense, white fumes start to appear, remove the heat and allow acid to cool. Store the concentrated acid in a glass container.

   d. Application.
      (1) *General.* Commercial sulfuric acid is available in 13 gallon carboys. Smaller quantities of this acid are available in chemical laboratory reagent storage containers. It is recommended that a small quantity of acid, about one pint, be secured and stored in a glass container until it is used.

      (2) *Use with delay mechanisms.*
         (*a*) Construction of specific delay mechanisms is described in chapter 5. Within the delay mechanism, there is a container filled with acid. The acid corrodes this container, is absorbed by the container material or is spilled from the container until it comes in contact with the igniter mixture.

         (*b*) Carefully fill the container in the delay mechanism with concentrated sulfuric acid. This can be accomplished easily with a small glass funnel. A medicine dropper is used when the delay mechanism container is small.

Caution: **Concentrated sulfuric acid must be handled carefully because it is very corrosive. If it is splashed on clothing, skin or eyes, the affected area must be immediately flushed with water. This may not be always practical. It is recommended that eye protection be worn by the user when pouring concentrated sulfuric acid. Many types are** available for this purpose. Rubber gloves can be worn to protect the hands. A small bottle of water can be carried to flush small areas of skin or clothing which may be contaminated with the acid.

(3) *Manual application.*

    (a) Manual application of concentrated sulfuric acid for direct initiation of an igniter is not recommended when fuse cord is available. It is possible to employ this acid for direct initiation by quickly adding three or four drops to the igniter material. This can be accomplished with a medicine dropper. Keep hands and clothing clear of the igniter; ignition may take place almost instantly with addition of acid.

Caution: Do not allow material such as sugar, wood, cotton or woolen fibers to fall into the *boiling acid.* A violent reaction could occur with splattering of acid.

    (b) Since sulfuric acid has a unique freezing point related to acid concentration, the information shown below is useful when this acid is used with delay mechanisms in low temperature surroundings. Be sure of acid concentration by checking with a hydrometer.

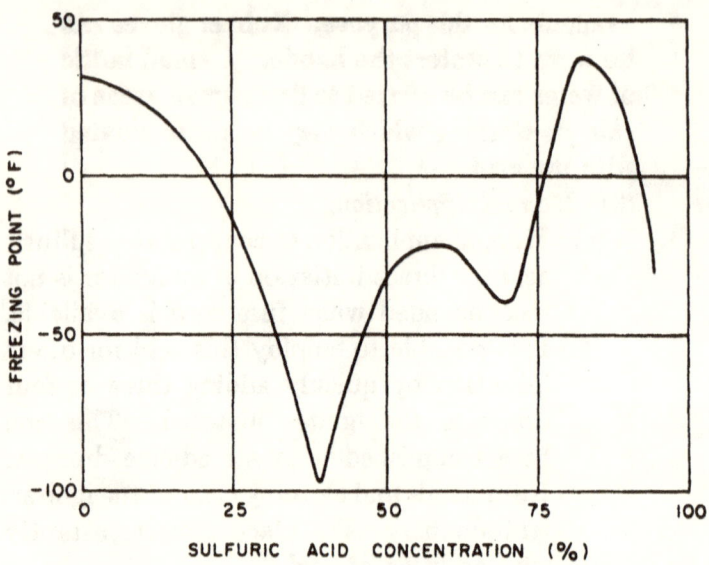

Sulfuric acid concentration (%)	Specific gravity	Freezing point (° F.)
0	1.000	+32
10	1.074	+23
20	1.151	+5
30	1.229	−39
39	1.295	−97
40	1.306	−91
50	1.408	−31
60	1.510	−22
70	1.611	−40
75	1.686	−7
77	1.706	+12
80	1.726	+27
81	1.747	+39
89	1.818	+24
90	1.824	+13
92	1.830	−1
93	1.835	−29

0104. WATER

a. *Description.*
   (1) Water causes spontaneous initiation of certain igniter mixtures. This is caused by a chemical reaction of the igniter materials in the presence of water. The following igniters are initiated by water: Sugar—Sodium Peroxide (0203), Aluminum Powder—Sodium Peroxide (0204), and Silver Nitrate—Magnesium Powder (0208).
   (2) The most important use for water as an initiator is in conjunction with delay mechanisms. Since only a few igniter mixtures are initiated by water and *it cannot be used at freezing temperatures*, its use is limited. When tactics so dictate, water can be reliably used with the following delay mechanisms: Gelatin Capsule (0402), Overflow (0407), Tipping Delay—Filled Tube (0408), Balancing Stick (0410), and Stretched Rubber Band (0411).

   Note. Sulfuric acid of any concentration can be substituted for water in the initiation of water activated igniters. Water *cannot* be substituted for concentrated sulfuric acid.

b. *Material and Equipment.* Water.
c. *Preparation.* None.
d. *Application.*
   (1) *Use with delay mechanisms.* Construction of specific delay mechanisms is presented in chapter 5. Within the delay mechanism, there is a container filled with water. The water dissolves the container or is spilled from the container and comes in contact with igniter mixture, initiating the fire train.

(2) *Manual application.* Fuse cord, when available, is recommended in preference to water as an initiator. Water is used for direct ignition of a specific igniter by adding drops as with a medicine dropper. Keep hands and clothing clear of the igniter; ignition may take place almost instantaneously with addition of water.

CHAPTER 3
IGNITERS

0201. SUGAR-CHLORATE

  a. *Description.*
   (1) This item consists of a mixture of granulated sugar and potassium chlorate or sodium chlorate. It can be used to ignite all the incendiaries listed in chapter 4 except Thermite (0307). It may be used directly as an incendiary on readily flammable material such as rags, dry paper, dry hay, or in the combustible vapor above liquid fuels.
   (2) The igniter can be initiated by Fuse Cord (0101), Improvised String Fuse (0102), or Concentrated Sulfuric Acid (0103).
   (3) This simple sugar-chlorate mixture closely resembles granulated sugar and should not ordinarily arouse suspicion. It is an excellent igniter.

  Caution: **This mixture is poisonous and must not be eaten.**

  b. *Material and Equipment.*
   Granulated sugar (do *not* use powdered or confectioners sugar.)
   Potassium chlorate or sodium chlorate (no coarser than granulated sugar).
   Spoon (preferably nonmetallic).
   Container with tight-fitting lid.
   Rolling pin or round stick.

c. *Preparation.*
   (1) Using a clean, dry spoon, place granulated sugar in the container to one-quarter container volume. Wipe the spoon with a clean cloth.
   (2) If the potassium or sodium chlorate is lumpy, remove all lumps by crushing with a rolling pin. Using the spoon, add an equal quantity of chlorate to the container.

Caution: **If this mixture is carelessly handled with excessive bumping and scraping, it could be a fire hazard to the user. As a precaution, remove any mixture adhering to the lip or edge of the jar before tightening the lid.**

   (3) Tighten the lid of the jar, turn the jar on its side and slowly roll until the two powders are completely mixed. The mixture is now ready for use. It may be stored for months in a tightly sealed container.

d. *Application.*
   (1) Carefully pour or spoon the mixture, in a single pile, on the incendiary. Prepare the mixture for ignition with Fuse Cord (0101) or Improvised String Fuse (0102) in the normal manner. The fuse cord should terminate near the center of the igniter mixture. Concentrated Sulfuric Acid (0103) can be used as an initiator, but is generally less convenient. Ignition takes place almost immediately on contact with the acid. Acid is recommended for use with specific delay mechanisms found in chapter 5.
   (2) If only battery-grade sulfuric acid is available, it must be concentrated before use to a

specific gravity of 1.835 by heating it in an enameled, heat-resistant glass or porcelain pot until dense, white fumes start to appear. See paragraph 0103 for details.

(3) When used to ignite flammable liquids, wrap a quantity of the mixture in a nonabsorbent material and suspend it inside the container near the open top. The container must remain open for easy ignition and combustion of the flammable liquid.

(4) To minimize the hazard of premature ignition of flammable liquid vapors, allow at least two feet of fuse length to extend from the top edge of an open container of flammable liquid before lighting the fuse.

0202. FIRE FUDGE

a. Description.

(1) This item consists of a mixture of sugar and potassium chlorate in a hot water solution which solidifies when cooled to room temperature. It can be used to ignite all the incendiaries listed in chapter 4 except Thermite (0307). It may be used directly as an incendiary on readily flammable material, such as rags, dry paper, dry hay, or in the combustible vapor above liquid fuels.

(2) The igniter can be initiated by Fuse Cord (0101), Improvised String Fuse (0102), or Concentrated Sulfuric Acid (0103).

(3) Fire fudge resembles a white sugar fudge having a smooth, hard surface. The advantage of this igniter material over Sugar-Chlorate (0201), is its moldability. The

procedure for preparation must be followed closely to obtain a smooth, uniform material with a hard surface.

Caution: **This material is poisonous and must not be eaten.**

b. *Material and Equipment.*

Granulated sugar (do *not* use powdered or confectioners sugar).

Potassium chlorate (no coarser than granulated sugar).

Metallic, glass or enameled pan.

Measuring container (size of this container determines quantity of finished product).

Spoon (preferably nonmetallic).

Thermometer (to read in the range 200° F. to 250° F.)

Heat source.

c. *Preparation.*

(1) Clean the pan by boiling some clean water in it for about five minutes. Discard the water, pour one measureful of clean water into the pan and warm it. Dry the measuring container and add one measureful of sugar. Stir the liquid until the sugar dissolves.

(2) Boil the solution until a fairly thick syrup is obtained.

(3) Remove the pan from the source of heat to a distance of at least six feet and shut off heat. Rapidly add two measurefuls of potassium chlorate. Stir gently for a minute to mix the syrup and powder, then pour or spoon the mixture into appropriate molds. If the mold is paper, it can usually be peeled off when the fire fudge cools and hardens. Pieces of card-

board or paper adhering to the igniter will not impair its use. Pyrex glass or ceramic molds can be used when a clear, smooth surface if desired. It is recommended that section thickness of molded fire fudge be at least one-half inch. If desired, molded fire fudge can be safely broken with the fingers.

(4) This material is moderately hard immediately after cooling. It will become harder after 24 hours. When kept in a tightly sealed container, it will retain its effectiveness for months.

Caution: If this igniter material is carelessly handled with excessive bumping or scraping, it could be a fire hazard to the user.

d. *Application.*

(1) Place a piece of fire fudge on top of the incendiary. Minimum size should be about one inch square and one-half inch thick Prepare the fire fudge for ignition with Fuse Cord (0101) or Improvised String Fuse (0102) in the normal manner. Concentrated Sulfuric Acid (0103) can be used as an initiator but is generally less convenient. Acid is recommended for use with specific delay mechanisms found in chapter 5.

(2) If only battery-grade sulfuric acid is available it must be concentrated before use to a specific gravity of 1.835 by heating it in an enameled, heat resistant glass or porcelain pot until dense, white fumes start to appear. See paragraph 0103 for details.

(3) When used to ignite flammable liquids, wrap a quantity of the igniter mixture in a nonabsorbent material and suspend it inside the container near the open top. The container must remain open for easy ignition and combustion of the flammable liquid.

(4) To minimize the hazard of premature ignition of flammable liquid vapors, allow at least two feet of fuse length to extend from the top edge of an open container of flammable liquid before lighting the fuse.

0203. SUGAR—SODIUM PEROXIDE

a. *Description.*

(1) This item consists of a mixture of sodium peroxide and granulated sugar. It can be used to ignite all the incendiaries listed in chapter 4 except Thermite (0307). It may be used directly as an incendiary on readily flammable material such as rags, dry paper dry hay, or in the combustible vapor above liquid fuels.

(2) The igniter can be initiated by Fuse Cord (0101), Improvised String Fuse (0102), Concentrated Sulfuric Acid (0103), or Water (0104).

Caution: **This mixture is unstable and can ignite at high humidity or when wet slightly by drops of water, perspiration, etc.**

b. *Material and Equipment.*

Granulated sugar (do *not* use powdered or confectioners sugar).

Sodium peroxide (no coarser than granulated sugar).

Spoon.
Container with tight fitting lid for mixing and storage.
c. *Preparation.*
  (1) Using a clean, dry spoon, place granulated sugar in the container to one-quarter container volume.
  (2) Wipe the spoon with a clean, dry cloth, and add an equal amount of sodium peroxide to the dry mixing container. Tighten the lid on the sodium peroxide container, and remove it at least six feet from the working area.
  (3) Tighten the lid on the mixing container. Turn the container on its side and slowly roll until the two powders are completely mixed. The mixture is now ready for use.
  (4) A good practice is to keep the granulated sugar and sodium peroxide in separate airtight containers and mix just before use.

Caution: **Do not store this mixture longer than three days because decomposition may occur and cause spontaneous combustion. Be sure that the storage container is air-tight.**

d. *Application.*
  (1) Carefully pour or spoon the mixture, in a single pile, on the incendiary. Prepare the mixture for ignition with Fuse Cord (0101) or Improvised String Fuse (0102) in the normal manner. The fuse cord should terminate near the center of the igniter mixture. Concentrated Sulfuric Acid (0103) and Water (0104) can be used as initiators, but are generally less convenient. Ignition takes place almost immediately on contact with the

acid or water. These liquid initiators are convenient for use with specific delay mechanisms found in chapter 5.

(2) When used to ignite flammable liquids, wrap a quantity of the mixture in a non-absorbent material and suspend it inside the container near the open top. The container must remain open for easy ignition and combustion of the flammable liquid.

(3) To minimize the hazard of premature ignition of flammable liquid vapors, allow at least two feet of fuse length to extend from the top of an open container of flammable liquid before lighting the fuse.

0204. ALUMINUM POWDER—SODIUM PEROXIDE

a. Description.

(1) This item consists of a mixture of sodium peroxide and powdered aluminum. It can be used to ignite all the incendiaries listed in chapter 4 except Thermite (0307). It may be used directly as an incendiary on readily flammable material, such as rags, dry paper, dry hay or in the combustible vapor above liquid fuels.

(2) The igniter can be initiated by Fuse Cord (0101), Improvised String Fuse (0102), Concentrated Sulfuric Acid (0103), or water (0104).

Caution: **This mixture is unstable and can ignite at high humidity or when wet slightly by drops of water, perspiration, etc.**

b. *Material and Equipment.*

Powdered aluminum (no coarser than granulated sugar).

Sodium peroxide no coarser than granulated sugar).

Spoon.

Container with tight fitting lid for mixing and storage.

c. *Preparation.*
  (1) Using a clean, dry spoon, place powdered aluminum in the container to one-quarter container volume.
  (2) Wipe the spoon with a clean, dry cloth, and add an equal amount of sodium peroxide to the dry mixing container. Tighten the lid on the sodium peroxide container, and remove it at least six feet from the working area.
  (3) Tighten the lid of the mixing container. Turn the container on its side and slowly roll until the two powders are completely mixed. The mixture is now ready to use.
  (4) A good practice is to keep the powdered aluminum and sodium peroxide in separate containers and mix just before use.

  Caution: **Do not store this mixture longer than three days because decomposition may occur and cause spontaneous combustion. Be sure that the storage container is air-tight.**

d. *Application.*
  (1) Carefully pour or spoon the mixture, in a single pile, on the incendiary. Prepare the mixture for ignition with Fuse Cord (0101) or Improvised String Fuse (0102) in the normal manner. The fuse cord should terminate

near the center of the igniter mixture. Concentrated Sulfuric Acid (0103) and Water (0104) can be used as initiators, but are generally less convenient. Ignition takes place almost immediately on contact with the acid or water. These liquid initiators are convenient for use with specific delay mechanisms found in (chapter 5.)

(2) When used to ignite flammable liquids, wrap a quantity of the mixture in a nonabsorbent material and suspend it inside the container near the open top. The container must remain open for easy ignition and combustion of the flammable liquid.

(3) To minimize the hazard of premature ignition of flammable liquid vapors, allow at least two feet of fuse length to extend from the top edge of an open container of flammable liquid before lighting the fuse.

0205. MATCH HEAD

a. Description.

(1) This item consists of a quantity of match heads, prepared by breaking the heads off their match sticks and grouping the match heads together to form the desired quantity of igniter. Any kind of friction match will do. It can be used to ignite the following incendiaries listed in chapter 4: Napalm (0301), Gelled Gasoline (exotic thickeners) (0302), Gelled Gasoline (improvised thickeners) (0303), Paraffin-Sawdust (0304), and Flammable Liquids (0308). It may be used directly

as an incendiary on readily flammable material such as rags, dry paper, dry hay or in the combustible vapor above liquid fuels.

(2) The igniter can be initiated by a match flame, Fuse Cord (0101), Improvised String Fuse (0102), or Concentrated Sulfuric Acid (0103).

b. *Material and Equipment.*
Razor blade or knife.
Container with tight-fitting lid.
Matches, friction.

c. *Preparation.*
(1) Using a knife or razor blade, cut off the match heads.
(2) Prepare the desired quantity of igniter and store it in an airtight container until ready for use.

d. *Application.*
(1) Pour or spoon the match heads, in a single pile, on the incendiary. Prepare the match heads for ignition with Fuse Cord (0101) or Improvised String Fuse (0102) in the normal manner. The fuse cord should terminate near the center of the match head pile. Concentrated Sulfuric Acid (0103) or a match flame can also be used as an initiator. Ignition takes place almost immediately on contact with the acid or the match flame. Acid is recommended for use with specific delay mechanisms found in chapter 5.
(2) If only battery-grade sulfuric acid is available, it must be concentrated before use to a specific gravity of 1.835 by heating it in an enameled, heat-resistant glass or porcelain

pot until dense, white fumes start to appear. See paragraph 0103 for details.

(3) When used to ignite flammable liquids, wrap a quantity of the match heads in a nonabsorbent material and suspend it inside the container near the open top. The container must remain open for easy ignition and combustion of the flammable liquid.

(4) To minimize the hazard of premature ignition of flammable liquid vapors, allow at least two feet of fuse length to extend from the top edge of an open container of flammable liquid before lighting the fuse.

0206. POTASSIUM PERMANGANATE—GLYCERIN

a. Description.

(1) This item consists of a small pile of potassium permanganate crystals which are ignited by the chemical action of glycerin on the crystals. It can be used to ignite all the incendiaries listed in chapter 4 except Thermite (0307). It may be used directly as an incendiary on readily flammable material, such as rags, dry paper, dry hay, or in the combustible vapor above liquid fuels.

(2) Ignition is accomplished by causing a few drops of glycerin to contact the potassium permanganate crystals. A hotter flame is produced when powdered magnesium or powdered aluminum is mixed with the the potassium permanganate crystals.

(3) Ignition time, after addition of the glycerin, increases as temperature decreases. This igniter is not reliable below 50° F.

b. *Material and Equipment.*
Potassium permanganate crystals (no coarser than granulated sugar).
Glycerin.
One small container with tight-fitting lid for the glycerin.
One larger container with tight-fitting lid for the potassium permanganate crystals.
Powdered magnesium or powdered aluminum (no coarser than granulated sugar).
Preparation.
(1) Put some glycerin in the small container and cap tightly.
(2) Fill the larger container with potassium permanganate crystals and cap tightly.
(3) If powdered magnesium or powdered aluminum is available, mix 85 parts potassium permanganate crystals and 15 parts powdered magnesium or powdered aluminum and store this mixture in the large bottle.
(4) Keep these containers tightly sealed and the material in the containers will remain effective for a long period of time.

d. *Application.* Pour out a quantity of the potassium permanganate crystals (with or without powdered aluminum or powdered magnesium), in a single pile on the incendiary. Manual ignition is accomplished by causing a few drops of glycerin from a medicine dropper to come in contact with the potassium permanganate crystals. Keep hands and clothing clear of the igniter; ignition may take place almost instantly with addition of the glycerin. This igniter is convenient for use with specific delay mechanisms found in chapter 5.

0207. POWDERED ALUMINUM—SULFUR PELLETS

a. Description.

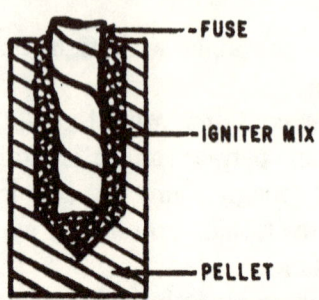

(1) This igniter consists of finely powdered aluminum, sulfur and starch which have been thoroughly mixed and shaped into hardened cylindrical pellets. It can be used to ignite all the incendiaries listed in chapter 4. It is an excellent igniter for Thermite (0307). It may be used directly as an incendiary on readily flammable material such as rags, dry paper, dry hay, or in the combustible vapor above liquid fuels.

(2) The igniter can be initiated by Fuse Cord (0101) or Improvised String Fuse (0102). A hole is made in one pellet to receive a fuse and a small quantity of another more easily started igniter mixture. A number of unmodified pellets are attached to the first pellet to increase the quantity of heat after combustion occurs.

b. Material and Equipment.

Finely powdered aluminum (no coarser than cake flour).

Finely powdered sulfur (no coarser than cake flour).
Finely powdered starch (no coarser than cake flour).
Water.
Cylindrical tube about 4 inches long and 3/4 inch inside diameter made of metal, wood, glass or plastic.
Rod which fits into the above tube.
Rod about 3/8 inch in diameter (should be about one-half the inside diameter of the 4-inch long tube).
Mixing bowl.
Tablespoon.
Teaspoon.
Stove or hot plate.
Knife.
Measuring container.

c. *Preparation.*

(1) Place six tablespoons of aluminum powder in a mixing bowl then add 15 tablespoons of powdered sulfur.

(2) Mix the two powders gently with the spoon for a few minutes until no unmixed particles of sulfur are visible.

(3) In a separate pot add two teaspoons of laundry starch to about 6 ounces of water and boil gently for a few minutes. Stir until the starch is dissolved and allow the solution to cool to room temperature.

(4) When cool, take about one-half of the starch solution and add it to the mixture of aluminum and sulfur powder.

(5) Mix with a spoon until the whole mass is a smooth, evenly mixed, putty-like paste.

(6) Fill the cylindrical tube with this paste, place one end of this tube on a hard surface and tamp the paste with the 3/8 inch diameter rod to squeeze out the air bubbles and consolidate the paste.

(7) Push the paste out of the tube with the larger rod, which just fits the tube, so that it forms a cylinder, then cut the damp cylinder into 1½ inch lengths using the knife.

(8) Dry these pieces at 90° F. for at least 24 hours before using. The drying time can be reduced by using a drying oven at a maximum temperature of 150° F.

(9) Form a hole at least ½ inch in diameter approximately half-way into one end of an igniter pellet.

(10) Put one of the following igniters into the cavity to roughly one-half its depth:
Sugar-Chlorate (0201)
Sugar—Sodium Peroxide (0203)
Aluminum Powder—Sodium Peroxide (0204)
Silver Nitrate—Magnesium Powder (0208)

(11) Insert a length of fuse into the hole so that it makes contact with the igniter mix. Fill the remainder of the hole with igniter mix and tamp down to hold the fuse firmly.

(12) Tape the fuse cord in place to prevent it from working loose and falling out.

(13) Tape two or more pellets without holes to the one with the fuse.

(14) Store all the pellets in a dry, closed container until required for use.

d. *Application.*

(1) For ignition of thermite, a cluster of at least three pellets should be used. Bury the cluster of igniter pellets just below the surface of the thermite, with the fuse extending for ignition by a match flame. Large quantities of thermite may require a cluster of more than three pellets for satisfactory ignition.

(2) For use as an igniter of a solid incendiary, place a cluster of pellets on top of the incendiary.

(3) When used to ignite flammable liquids, wrap a cluster of igniter pellets in a nonabsorbent material and suspend it inside the container near the open top. The container must remain open for easy ignition and combustion of the flammable liquid.

(4) To minimize the hazard of premature ignition of flammable liquid vapors, allow at least two feet of fuse length to extend from the top edge of an open container of flammable liquid before lighting the fuse.

0208. SILVER NITRATE—MAGNESIUM POWDER

a. *Description.*

(1) This item consists of a mixture of silver nitrate crystals and magnesium powder. It can be used to ignite all the incendiaries listed in chapter 4 except Thermite (0307). It may be used directly as an incendiary on readily flammable material such as rags, dry paper, dry hay, or in the combustible vapor above liquid fuels.

(2) The igniter can be initiated by Fuse Cord (0101), Improvised String Fuse (0102), Concentrated Sulfuric Acid (0103), or Water (0104).

Caution: **This mixture is unstable and may ignite at high humidity or when wet slightly by drops of water, perspiration, etc.**

b. *Material and Equipment.*
Silver nitrate crystals (no coarser than granulated sugar).
Magnesium powder or filings (no coarser than granulated sugar).
Spoon.
Container with tight-fitting lid.

c. *Preparation.*
(1) Using a clean, dry spoon, place magnesium powder or filings into the dry mixing container to one-quarter container volume. If magnesium filings are used, they should be free of grease.
(2) Wipe the spoon with a clean, dry cloth, then add an equal quantity of silver nitrate crystals to the dry mixing container. Tighten the lid on the silver nitrate container, and remove it at least six feet from the working area.
(3) Tightly close the lid on the mixing container. Turn the container on its side and slowly roll until the two powders are completely mixed. The mixture is now ready for use.
(4) A good practice is to keep the silver nitrate crystals and the magnesium powder or filings in separate air-tight containers and mix just before use.

Caution: This mixture should be kept out of direct sunlight to avoid decomposition of the silver nitrate which could render this igniter mixture ineffective.

d. *Application.*

(1) Carefully pour or spoon the mixture, in a single pile, on the incendiary. Prepare the mixture for ignition with either Fuse Cord (0101) or Improvised String Fuse (0102) in the normal manner. The fuse cord should terminate near the center of the igniter mixture. Concentrated Sulfuric Acid (0103) and Water (0104) can be used as initiators but are generally less convenient. Ignition takes place almost immediately on contact with the acid or water. These liquid initiators are convenient for use with specific delay mechanisms found in chapter 5.

(2) When used to ignite flammable liquids, wrap a quantity of the mixture in a nonabsorbent material and suspend it inside the container near the open top. The container must remain open for easy ignition and combustion of the flammable liquid.

(3) To minimize the hazard of premature ignition of flammable liquid vapors, allow at least two feet of fuse length to extend from the top edge of an open container of flammable liquid before lighting the fuse.

0209. WHITE PHOSPHORUS

a. *Description.*

(1) This item consists of white phosphorus dissolved in carbon disulfide. It can be used to

ignite the following incendiaries listed in chapter 4: Napalm (0301), Gelled Gasoline (exotic thickeners) (0302), Gelled Gasoline (improvised thickeners) (0303), and Paraffin-Sawdust (0304). It may be used directly as an incendiary on readily flammable material such as rags, dry paper, dry hay, or in the combustible vapor above liquid fuels.

(2) Ignition is achieved when the volatile solvent, carbon disulfide, evaporates and the white phosphorus comes in contact with air.

Caution: **Never touch white phosphorus directly or allow any of its solutions to touch the skin. Painful burns which heal very slowly may result. White phosphorus sticks must *always* be stored completely under water. If any of the phosphorus solution is accidently spilled on the skin, immediately flush the affected area with water; then decontaminate the affected area by dabbing with copper sulfate solution.**

b. *Material and Equipment.*

White phosphorus sticks (sometimes called yellow phosphorus).

Carbon disulfide.

Copper sulfate solution.

Tweezers or tongs.

Two glass containers about 8-ounce capacity with lids or stoppers made of glass, earthenware, or metal. Do not use a rubber lid or stopper (carbon disulfide will attack rubber).

c. *Preparation.*

(1) Prepare some copper sulfate solution by adding one spoonful of copper sulfate crystals

to one of the glass containers. Fill the container with water, place the stopper in the open mouth of the bottle and shake until the crystals dissolve.

(2) Pour carbon disulfide into the other glass container to one-quarter container volume.

Caution: **Carbon disulfide fumes are poisonous. Always cap an open container of carbon disulfide as soon as possible. Work in a well ventilated area.**

(3) With a pair of tweezers remove some sticks of white phosphorus from their storage container. Totally submerge them immediately in the carbon disulfide to bring the level up to one-half full. Be sure that all the phosphorus left in the original container is completely submerged in water before putting the container away. Wash the tweezers immediately in the copper sulfate solution.

(4) Securely stopper the bottle containing the white phosphorus and carbon disulfide and allow to stand until the white phosphorus dissolves. This usually takes about eight hours. The time required to dissolve white phosphorus can be reduced by shaking the bottle. Be sure that the bottle top does not come off.

(5) Do not store in direct sunlight because the solution will become ineffective. This solution should never be stored more than three days.

Note. If carbon disulfide is not available, benzene (benzol) may be used to dissolve the phosphorus. It requires considerable shaking and overnight soaking to get an appreciable amount of phosphorus dissolved in benzene. *Do not* attempt to use red phosphorus for preparing this igniter because it does not behave like white phosphorus.

d. Application.

(1) To ignite readily flammable material, pour the white phosphorus solution directly onto the material; it will ignite when the solvent evaporates, exposing the white phosphorus to the air. Once the solution is poured, the empty bottle should be discarded immediately because any solution remaining on the bottle will ignite when the solvent evaporates. Do not cover the soaked flammable material because the carbon disulfide must evaporate for ignition to occur.

(2) The incendiaries mentioned under *Description* above can be initiated by first impregnating crumpled paper or absorbent paper towels with the white phosphorus solution and placing the impregnated paper on the material to be ignited.

(3) Delay times of the phosphorus solution may be varied by the addition of gasoline or toluene (toluol). Add a small quantity of either solvent to the original white phosphorus solution and test the solution each time until the desired delay time is achieved. Delay times of 20 to 30 minutes may be obtained in this manner.

(4) Check the delay time under conditions expected at the target. Air currents hasten the evaporation of the solvent and decrease delay time. A high ambient temperature will also decrease delay time whereas a low ambient

temperature will increase the delay time. This igniter is not reliable at or below freezing temperatures (32° F.)

(5) To make incendiary paper, soak strips of ordinary writing paper in the phosphorus-carbon disulfide for a few minutes. Remove the paper with a pair of tweezers or tongs and place in a vial filled with water. Be sure to wash off the tweezers immediately in copper sulfide solution. Cap the vial and store until ready to use. To use this incendiary paper, remove the strips of paper with a pair of tweezers, and place among the material to be ignited.

0210. MAGNESIUM POWDER—BARIUM PEROXIDE

a. Description.

(1) This item consists of a mixture of finely powdered magnesium and finely powdered barium peroxide. It can be used to ignite all the incendiaries listed in chapter 4 and is particularly suited for ignition of thermite. It may be used directly as an incendiary on readily flammable material such as rags, dry paper, dry hay, or in the combustible vapor above liquid fuels.

(2) The igniter can be initiated by Fuse Cord (0101) or Improvised String Fuse (0102).

b. Material and Equipment.

Magnesium powder (no coarser than table salt).
Barium peroxide (no coarser than table salt).
Spoon.
Container with tight-fitting lid.

c. *Preparation.*
   (1) Using a clean, dry spoon, place powdered magnesium into the dry mixing container to one-quarter container volume.
   (2) Wipe the spoon with a clean, dry cloth, then add powdered barium peroxide to the dry mixing container to three-quarters container volume. Tighten the lid on the barium peroxide container, and remove it at least six feet from the working area.
   (3) Tightly close the lid on the mixing container. Turn the container on its side and slowly roll until the two powders are completely mixed. The mixture is now ready for use.
   (4) A good practice is to keep the powdered magnesium and powdered barium peroxide in separate containers and mix just before use.

d. *Application.*
   (1) Carefully pour or spoon the mixture, in a single pile, onto the incendiary. Prepare the mixture for ignition with Fuse Cord (0101) or Improvised String Fuse (0102) in the normal manner. The fuse cord should terminate near the center of the igniter mixture.
   (2) In ignition of thermite, spread the igniter mixture to a depth of at least $\frac{1}{4}$ inch on the top surface of the thermite which is held in an assembly described under *Application* of Thermite incendiary (0307). The fuse cord will initiate the thermite igniter which will, in turn, ignite the thermite.
   (3) When used to ignite flammable liquids, wrap a quantity of the mixture in a nonabsorbent material and suspend it inside the container

near the open top. The container must remain open for easy ignition and combustion of the flammable liquid.

(4) To minimize the hazard of premature ignition of flammable liquid vapors, allow at least two feet of fuse length to extend from the top edge of an open container of flammable liquid before lighting the fuse.

0211. SUBIGNITER FOR THERMITE

a. Description.

(1) This item consists of a mixture of a metal powder and an oxidizing agent. Two metal powder alternates and four oxidizing agent alternates are specified. In the combustion process, the metal powder is oxidized, resulting in the liberation of a large quantity of heat.

(2) This subigniter is a substitute for Magnesium Powder--Barium Peroxide Igniter (0210), and should be used only if that Igniter is not available. The disadvantage of this subigniter is that it cannot be directly initiated by fuse cord. To use this subigniter for initiating thermite, it is necessary to use another igniter mixture to initiate the subigniter, preferably Sugar-Chlorate (201). The fuse cord will initiate the sugar-chlorate, which will, in turn, ignite the subigniter and, thereby, initiate the thermite.

(3) This subigniter can be directly initiated by all the igniters listed in chapter 3 except White Phosphorus (0209).

b. Material and Equipment.
   Either aluminum or magnesium filings or powder (no coarser than granulated sugar).
   Any one of the following oxidizing agents: sodium dichromate, potassium permanganate, potassium nitrate, or potassium dichromate (no coarser than granulated sugar).
   Container with tight-fitting lid.
c. Preparation.
   (1) Using a clean, dry spoon, place one of the metal powders or filings in the container to one-third container volume. If metal filings are used, they should be free of grease.
   (2) Wipe the spoon with a clean, dry cloth and add an equal quantity of one of the above oxidizing agents.
   (3) Tighten the lid on the mixing container, turn the container on its side and slowly roll until the two powders are completely mixed. The mixture is now ready to use and may be stored for months in this tightly sealed container.
d. Application.
   (1) To use this subigniter, spread the material to a depth of at least 1/4 inch on the top surface of the thermite which is held in an assembly described under *Application* of Thermite Incendiary (0307). Spread another igniter, preferably Sugar-Chlorate (0201) on top of this subigniter to about the same depth. Prepare the mixture for ignition with Fuse Cord (0101) or Improvised String Fuse (0102) in the normal manner. The fuse cord should terminate near the center of the igniter

mixture. The fuse cord initiates the sugar-chlorate igniter which ignites the thermite subigniter which then ignites the thermite.

(2) For delay times longer than those conveniently obtained with fuse cord in ignition of thermite by this subigniter method, refer to chapter 5.

Caution: **Never attempt to ignite thermite subigniter without at least a few seconds delay fuse. It burns extremely fast and hot, and the user could be seriously burned if he were too close when ignition occurred.**

CHAPTER 4
INCENDIARY MATERIALS

0301. NAPALM

a. *Description.*

(1) This item consists of a liquid fuel which is gelled by the addition of soap powder or soap chips. It is easily prepared from readily available materials.

(2) This incendiary can be directly initiated by a match flame. However, if delay is required, the incendiary can be reliably initiated by a delay system consisting of any igniter listed in chapter 3 coupled with a delay mechanisms found in chapter 5.

(3) Napalm incendiary is easily ignited and long burning, and is suitable for setting fire to large wooden structures and other large combustible targets: It adheres to objects, even on vertical surfaces.

b. *Material and Equipment.*

Soap powder or chips (bar soap can be easily shaved or chipped). Detergents *cannot* be used.

Any of the following liquid hydrocarbon fuels: gasoline, fuel oil, diesel oil, kerosene, turpentine, benzol or benzene, toloul or toluene.

A double boiler made from any material with the upper pot having a capacity of at least two quarts.

A spoon or stick for stirring.
A source of heat such as a stove or hot plate.
A knife or grater if only bar soap is available.
An air-tight container.

?. Preparation.
  (1) Fill bottom of double boiler with water and heat until the water boils. Shut off source of heat.
  (2) Place upper pot on top of bottom pot and remove both containers to a point several feet from the heat source.
  (3) Pour soap chips or powder into the upper pot of the double boiler to one-quarter of pot volume.
  (4) Pour any one of the liquid hydrocarbon fuels listed under *Material and Equipment* above into the upper pot containing the soap chips or powder until the pot is one-half full.

  Caution: **Keep these fuels away from open flames.**

  (5) Stir the mixture with a stick or spoon until it thickens to a paste having the consistency of jam. Do this in a well ventilated room where the vapors will not concentrate and burn or explode from a flame or spark.
  (6) If the mixture has not thickened enough after about 15 minutes of stirring, remove the upper pot and put it several feet from the heat source. Again bring the water in the lower pot to a boil. Shut off heat source, place upper pot in lower pot at a location several feet from the heat source and repeat stirring until the napalm reaches the recommended consistency.

(7) When the proper consistency is obtained, store the finished napalm in a tightly sealed container until used. Napalm will keep for months when stored this way.

d. Application.

(1) To use napalm most effectively, it should be spread out over the surface of the material to be burned. This will start a large area burning at once. A match can be used to directly initiate this incendiary. A short time delay in ignition can be obtained by combining Fuse Cord (0101) or Improvised String Fuse (0102) and one of the igniter mixtures found in chapter 3. (For example, several spoonfuls of Sugar-Chlorate mixture (0201) are placed in a nonabsorbent wrapping material. Fuse cord is buried in the center of the igniter mixture and the fuse is taped to the wrapping material. This assembly is placed directly on the napalm. Delay times are determined by the length of fuse. Suitable delay mechanisms are given in chapter 5 for delay times longer than those practical with fuse cord.)

(2) Napalm made with the more volatile fuels should not ordinarily be used with a delay longer than one hour because the liquid fuel evaporates and this can reduce its effectiveness. In very hot weather, or if the napalm is exposed to the direct rays of the sun, it is recommended that napalm be made with fuel oil. In extremely cold weather, it is recommended that napalm be made with gasoline.

(3) The destructive effect of napalm is increased when charcoal is added. The charcoal will

readily ignite and the persistent fire from the charcoal will outlast the burning napalm. It is recommended that at least one quart of napalm be used to ignite heavy wooden structures and large wooden sections. A minimum of one-half quart is recommended for wooden structures of small cross section.

0302. GELLED GASOLINE (EXOTIC THICKENERS)

a. Description.

(1) This item consists of gasoline which is gelled with small quantities of organic chemicals. The operation is carried out quickly, without heat, by addition of the chemicals while stirring.

(2) This incendiary can be directly initiated by a match flame. However, any igniter listed in chapter 3 can be used in conjunction with specific delay mechanisms found in chapter 5 for delayed ignition of this incendiary.

(3) Gelled gasoline incendiary is readily ignited, long burning, and is suitable for setting fire to large wooden structures and other large combustible targets. It adheres to objects, even on vertical surfaces.

b. Material and Equipment.

Gasoline.
Balance or scale.
Spoon or stick for stirring.
Large air-tight container.
Small jar.

One of the following seven additive systems:

Components	Grams added per gal gas	Trade name	Manufacturer
SYSTEM 1			
A.. Lauryl amine..	55..	Alamine 4D Formonyte 616 Armeen.	General Mills Foremost Chemical Armour Chemical.
B.. Toluene diisocyanate.	27..	Hylene TM-65 Nacconate 65.	DuPont National Aniline.
SYSTEM 2			
A.. Coco amine....	55..	Alamine 21 Formonyte 601.	General Mills Foremost Chemical.
B.. Toluene diisocyanate.	27..	Hylene TM-65 Nacconate 65.	DuPont National Aniline.
SYSTEM 3			
A.. Lauryl amine..	57..	Alamine 4D Formonyte 616 Armeen.	General Mills Foremost Chemical Armour Chemical.
B.. Hexamethylene diisocyanate.	25..	Hexamethylene diisocyanate.	Borden Chemical
SYSTEM 4			
A.. Oleyl amine....	59..	Alamine II Formonyte 608.	General Mills Foremost Chemical.
B.. Hexamethylene diisocyanate.	23..	Hexamethylene diisocyanate.	Borden Chemical
SYSTEM 5			
A.. t-Octyl amine..	51..	t-Octyl........	Rohm and Haas
B.. Toluene diisocyanate.	31..	Hylene TM-65 Nacconate 65.	DuPont National Aniline.

Components	Grams added per gal gas	Trade name	Manufacturer
SYSTEM 6			
A.. Coco amine	51..	Alamine 21 Formonyte 601.	General Mills Foremost Chemical.
B.. Naphthyl isocyanate.	31..	Naphthyl isocyanate.	Distillation Products Industry.
SYSTEM 7			
A.. Delta-aminobutylmethyldiethoxysilane.	51..	Delta silane	Union Carbide
B.. Hexamethylene diisocyanate.	31..	Hexamethylene diisocyanate.	Borden Chemical

c. *Preparation.*

(1) Determine the amount of gasoline to be gelled and place this amount in the large container.

Caution: **Keep this material away from open flames.**

(2) Weigh out the appropriate quantity of component A. This can be calculated by multiplying the number of gallons of gasoline by the figure given in the *Grams Added Per Gal. Gas.* column of systems. (For example, if System 1 is being used and five gallons of gasoline are being gelled, then (5×55) or 275 grams of Lauryl amine are required).

(3) Add component A to the gasoline and stir for a few minutes to dissolve.

Caution: **Both components A and B are corrosive to the skin. If any of these materials contact the skin, wash the area with detergent and water.**

    (4) Clean the small container used to weight component A thoroughly or use another container for weighing component B. Weigh out the proper quantity of component B. Calculate the proper amount as mentioned above for component A.

    (5) Stir the gasoline—component A mixture rapidly and add all of component B at once, not a little at a time. At the same time that component B enters the mixture, remove the stirring rod and allow a few minutes for the gelling to take place.

    (6) Store the gelled gasoline in a tightly sealed container until ready to use. It will keep for months when stored in this manner.

d. Application.

    (1) To use gelled gasoline most effectively, it should be spread out over the surface of the material to be burned. This will start a large area burning at once. A match can be used to directly initiate this incendiary. A short time delay in ignition can be obtained by combining Fuse Cord (0101), or Improvised String Fuse (0102) and one of the igniter mixtures found in chapter 3. (For example, several spoonfuls of Sugar-Chlorate Mixture (0201) are placed in a nonabsorbent wrapping material. Fuse cord is buried in the center of the igniter mixture and the fuse is taped to the wrapping material. This assembly is placed directly on the gelled

gasoline. Delay times are determined by the length of fuse. Suitable delay mechanisms are given in chapter 5 for delay times longer than those practical with fuse cord.)

(2) Gelled gasoline should not ordinarily be used with a delay longer than one hour because gasoline evaporates and this can reduce its effectiveness.

(3) The destructive effect of gelled gasoline is increased when charcoal is added. The charcoal will readily ignite and the persistent fire from the charcoal will outlast the burning gasoline. It is recommended that at least one quart of gelled gasoline be used to ignite heavy wooden structures and large wooden sections. A minimum of one-half quart is recommended for wooden structures of small cross section.

Note. All of the chemicals used for the gelling process *must* be added in a *liquid* state. Many of the chemicals solidify at near freezing temperatures (32° F.) and uniform gels are difficult to produce at these temperatures.

0303. GELLED GASOLINE (IMPROVISED THICKENERS)
0303.1 LYE SYSTEMS

(1) This item consists of gasoline which is gelled by the addition of certain ingredients that are locally available. The following eight basic systems will produce gelled gasoline and are easily prepared: Lye systems, Lye-alcohol systems, Lye-balsam systems, Soap-alcohol systems, Egg system, Latex system, Wax systems, and Animal blood systems. These systems are discussed in the subparagraphs under 0303.

(2) These incendiaries can be directly initiated by a match flame. However, any igniter listed in chapter 3 can be used in conjunction with specific delay mechanisms given in chapter 5 for delayed ignition.

(3) Gelled gasoline incendiary is readily ignited, long burning and is suitable for setting fire to large wooden structures and other large combustible targets. It adheres to objects, even on vertical surfaces.

b. *Material and Equipment.*

Ingredient	Parts by volume	Used for	Common source
Gasoline	60	Motor fuel	Gas stations or motor vehicles.
Lye	2 (flake) or 1 (powder)	Drain cleaner, making of soap, etc.	Food and drug stores, soap factories.
Water	1 or 2	(Always use about same amount as dry lye).	
Rosin powder.	15	Paint or varnish, Naval supply, industrial uses.	Food and drug stores, pine tree extract, pai..t and varnish factories.

Two air-tight containers
Spoon or stick for stirring

Note. Lye is also known as caustic soda or sodium hydroxide. Allow for strength of lye; if only 50% (as in Drano), use twice the amount indicated above. Castor oil can be substituted for the rosin. Potassium hydroxide (caustic potash, potassa) may be used in place of lye.

c. *Preparation.*

(1) Measure the required quantity of gasoline and place in a clean container.

Caution: **Keep material away from open flames.**

(2) Break the rosin into *small* pieces and add to the gasoline.

(3) Stir the mixture for about five minutes to disperse the rosin.

(4) In a separate container dissolve the lye in water.

Caution: **Add lye to water slowly. Do not prepare this solution in an aluminum container.**

(5) Add this solution to the gasoline mixture and stir until mixture thickens (about one minute).

(6) The mixture will thicken to a very firm butter paste within one to two days. The mixture can be thinned, if desired, by mixing in additional gasoline. Store in an air-tight container until ready to use.

d. Alternate Preparation Using Pyrethrum Extract Instead of Rosin.

(1) Replace rosin by the following:

Ingredient	Parts by volume	Used for	Common source
Pyrethrum extract (20%).	18	Insecticide, medicine.	Hardware stores, graden supply, drug stores.

(2) Measure 78 parts by volume of gasoline and place in a clean container.

Caution: **Keep material away from open flames.**

(3) Dissolve the pyrethrum extract in the gasoline by stirring.

(4) In another container dissolve the lye in water.

Caution: **Add lye to water slowly. Do not prepare this solution in an aluminum container.**

    (5) Add 4 parts by volume of the lye solution to the gasoline mixture.

    (6) Stir every 15 minutes until gel forms. Store in an air-tight container until ready to use.

e. Application

    (1) To use gelled gasoline most effectively, it should be spread out over the surface of the material to be burned. This will start a large area burning at once. A match can be used to directly initiate this incendiary. A short time delay in ignition can be obtained by combining Fuse Cord (0101) or Improvised String Fuse (0102) and one of the igniter mixtures found in chapter 3. (For example, several spoonfuls of Sugar-Chlorate Mixture (0201) are placed in a nonabsorbent wrapping material. Fuse cord is buried in the center of the igniter mixture and the fuse is taped to the wrapping material. This assembly is placed directly on the gelled gasoline. Delay times are determined by the length of fuse. Suitable delay mechanisms are given in chapter 5 for delay times longer than those practical with fuse cord.)

    (2) Gelled gasoline should not ordinarily be used with a delay longer than one hour because gasoline evaporates and this can reduce its effectiveness.

    (3) The destructive effect of gelled gasoline is increased when charcoal is added. The charcoal will readily ignite and the persistent

fire from the charcoal will outlast the burning gasoline. It is recommended that at least one quart of gelled gasoline be used to ignite heavy wooden structures and large wooden sections. A minimum of one-half quart is recommended for wooden structures of small cross section.

0303.2 LYE-ALCOHOL SYSTEMS

a. Description. See Paragraph 0303.1.

b. Material and Equipment.

Ingredient	Parts by volume	Used for	Common source
Gasoline	58	Motor fuel	Gas stations or motor vehicles.
Lye	2 (flake) or 1 (powder).	Drain cleaner, making of soap.	Food and drug stores, soap factories.
Water	1 or 2	(Always use about the same amount as dry lye).	
Ethyl alcohol.	3	Whiskey	Liquor stores.
Tallow	14	Food	Fat extracted from solid fat or suet of cattle, sheep, or horses.

Spoon or stick for stirring
Two air-tight containers

Note. Lye is also known as caustic soda or sodium hydroxide. Allow for strength of lye. If only 50% (as in Drano), use twice the amount indicated above. Methyl (wood) alcohol, isopropyl (rubbing) alcohol or antifreeze product can be substituted for whiskey, but their use produces softer gels. Potassium hydroxide (caustic potash, potassa) may be used in place of lye.

(1) The following can be substituted for the tallow in order of preference:
   (a) Wool grease (lanolin) (very good)—fat extracted from sheep wool.
   (b) Castor oil (good).
   (c) Any vegetable oil (corn, cottonseed, peanut, linseed, etc.).
   (d) Any fish oil.
   (e) Butter or oleo margarine.
(2) When using substitutes (1)(c) and (e) above, it will be necessary to double the recommended amount of fat and of the lye solution for satisfactory thickening.

c. *Preparation.*
   (1) Measure out the appropriate amount of gasoline and place in a clean container.
   Caution: Keep material away from open flames.
   (2) Add the tallow to the gasoline and stir for about one-half minute to dissolve the tallow.
   (3) Add the alcohol to the mixture.
   (4) In another container dissolve the lye in water.
   Caution: Add lye to water slowly. Do not prepare this solution in an aluminum container.
   (5) Add the lye solution to the gasoline mixture and stir occasionally until the mixture thickens (about one-half hour).
   (6) The mixture will thicken to a very firm butter paste in one to two days. The mixture can be thinned, if desired, by mixing in additional gasoline. Store in an air-tight container until ready to use.

d. *Application.* See paragraph 0303.1.

0303.3 LYE-BALSAM SYSTEMS

a. *Description.* See paragraph 0303.1.
b. *Material and Equipment.*

Ingredient	Parts by volume	Used for	Common source
Gasoline	80	Motor fuel	Gas stations or motor vehicles.
Either: Copaiba balsam Copaiba resin Jesuits' balsam.	14	Medicine, varnish, odor fixative.	Drug stores, varnish factories, perfume processors, natural oleoresin.
Or: Tolu balsam Tolu resin Thomas balsam.	14	Medicine, perfume, confectionery, fumigant, chewing gum.	Drug stores, perfume processors, candy manufacturers.
Lye	3	Drain cleaner, making of soap.	Food and drug stores, soap factories.
Water	3		

Spoon or stick for stirring
Two air-tight containers

Note. Lye is also known as caustic soda or sodium hydroxide. Allow for the strength of the lye. If only 50% (as in Drano), use twice the amount indicated above. Potassium hydroxide (caustic potash, potassa) may be used in place of lye.

c. *Preparation.*

(1) Dissolve the lye in water using a clean container.

Caution: **Add lye to water slowly. Do not prepare this solution in an aluminum container.**

TAGO 7189-C 63

(2) Stir gasoline and copaiba balsam in another clean container.

(3) Add the saturated lye solution to the gasoline mixture and stir until the gel forms. Store in an air-tight container until ready to use.

Note. Increase the lye solution to 10 parts by volume (5 parts lye, 5 parts water) if the gasoline does not thicken.

d. *Application.* See paragraph 0303.1.

0303.4 SOAP-ALCOHOL SYSTEMS

a. *Description.* See paragraph 0303.1.
b. *Material and Equipment.*

Ingredient	Parts by volume	Used for	Common source
Gasoline	36	Motor fuel	Gas stations or motor vehicles.
Ethyl alcohol.	1	Whiskey	Liquor stores
Laundry soap.	20 (powder) or 28 (flake).	Washing	Food stores

Air-tight container
Spoon or stick for stirring

Note. Methyl (wood) or isoprophyl (rubbing) alcohols can be substituted for the ethyl alcohol. When a stronger alcohol (150 proof) or one of the dry alcohol substitutes is used, add an amount of water to make the concentration 50% by volume. (The *percent* alcohol is equal to ½ of the *proof*—150 proof is 75% alcohol.)

(1) Unless the word *soap* actually appears somewhere on the container or wrapper (at retail store level), a washing compound may be assumed to be a synthetic detergent. Soaps

react with mineral salts in hard water to form a sticky insoluble scum while synthetic detergents do not. Synthetic detergents cannot be used.

(2) The following is a list of commercially available soap products (at retail store level):

Name	Manufacturer
Ivory Snow	Proctor and Gamble
Ivory Flakes	Proctor and Gamble
Lux Flakes	Lever Brothers
Chiffon Flakes	Armour
Palmolive Bar Soap	Colgate-Palmolive
Sweetheart Bar Soap	Manhattan Soap Company
Octagon Bar Soap	Colgate-Palmolive

(3) Home prepared bar soaps may be used in place of purchased bar soaps.

c. *Preparation.*

(1) Measure out the appropriate amount of gasoline and place in a clean container.

Caution: Keep material away from open flames.

(2) Add the alcohol to the gasoline.

(3) Add the soap powder to the gasoline-alcohol mixture, and stir occasionally until the mixture thickens (about 15 minutes).

(4) The mixture will thicken to a very firm butter paste in one to two days. It can be thinned, if desired, by mixing in additional gasoline. Store in an air-tight container until ready to use.

d. *Application.* See paragraph 0303.1.

0303.5 EGG SYSTEM

a. Description. See paragraph 0303.1.

b. Material and Equipment.

Ingredient	Parts by volume	Chemical name	Used for	Common source
Gasoline	85		Motor fuel	Gas stations or motor vehicles.
Egg whites (chicken, ostrich, duck, turtle, etc.).	14		Food, industrial processes.	Food stores, farms.

Use any one of the following additives:

Ingredient	Parts by volume	Chemical name	Used for	Common source
Table salt	1	Sodium chloride	Food, industrial processes.	Sea water, natural brine, food stores.
Ground coffee (not decaffeinized).	3		Beverage	Food stores, coffee processors.
Leaf tea	3		Beverage	Cacao trees, food stores.
Sugar	2	Sucrose	Sweetening foods, industrial processes.	Sugar cane, food stores.
Borax	2	Sodium tetraborate decahydrate.	Washing aid, industrial processes.	Natural in some areas, food stores.
Saltpeter (Niter).	1	Potassium nitrate.	Pyrotechnics, explosives, matches, medicine.	Natural deposits, drug stores.

Ingredient	Parts by volume	Chemical name	Used for	Common source
Epsom salts	1	Mganesium sulfate heptahydrate.	Medicine, mineral water, industrial processes.	Natural deposits, Kieserite, drug and food stores.
Washing soda (sal soda).	2	Sodium carbonate decahydrate.	Washing cleanser, medicine, photography.	Food, drug, and photo supply stores.
Baking soda	2	Sodium bicarbonate.	Baking effervescent salts, beverages, mineral waters, medicine, industrial processes.	Food and drug stores.
Aspirin (crushed).	2	Acetylsalicylic acid.	Medicine	Food and drug stores.

Spoon or stick for stirring
Two air-tight containers

c. *Preparation.*
  (1) Separate the egg white from the yolk as follows:
    (a) *Method 1.* Crack the egg at approximately the center. Allow the egg white to drain into a clean container. When most of the egg white has drained off, flip the yellow egg yolk from one-half shell to the other, each time allowing the egg white to drain into the container. Transfer the egg white to a capped jar for storage or directly into the container being used for the gelled flame

fuel. Discard the egg yolk. Repeat the process with each egg. Do not get the yellow egg yolk mixed into the egg white. If egg yolk gets into the egg white, discard the egg.

(b) *Method 2.* Crack the egg and transfer (CAREFULLY SO AS TO AVOID BREAKING THE YOLK) the egg to a shallow dish. Tilt the dish slowly and pour off the egg white into a suitable container while holding back the yellow egg yolk with a flat piece of wood, knife blade, or fingers. Transfer the egg white to a capped jar for storage or directly to the container being used for the gelled flame fuel. Discard the egg yolk. Repeat the process with each egg being careful not to get yellow egg yolk mixed in with the egg white. If egg yolk gets into egg white, discard the egg and wash the dish.

(2) Store egg white in an ice box, refrigerator, cave, cold running stream, or other cool area until ready to prepare the gelled flame fuel.

(3) Pour the egg white into a clean container.

(4) Add the gasoline.

Caution: **Keep material away from open flames.**

(5) Add the table salt (or one of its substitutes) and stir until the gel forms (about 5–10 minutes). Use within 24 hours. Thicker gelled flame fuels can be obtained by—

(a) Reducing the gasoline content to 80 parts by volume (NO LOWER); or

(b) Putting the capped jars in hot (65° C., 149° F.) water for ½ hour and then letting them cool tö ambient temperature. (DO NOT HEAT THE GELLED FUEL CONTAINING COFFEE.)

d. *Application.* See paragraph 0303.1.

0303.6 LATEX SYSTEM

a. *Description.* See paragraph 0303.1.

b. *Material and Equipment.*

Ingredient	*Parts by volume*	*Used for*	*Common source*
Gasoline	92	Motor fuel	Gas stations or motor vehicles.
Either:			
Latex commercial or natural.	7	Paints, adhesives, rubber products.	Natural from tree or plant, rubber cement, general stores.
Or:			
Guayule Gutta percha Balata.	7	Wire insulation, waterproofing, machinery belts, golf ball covers, gaskets.	Coagulated and dried latex, rubber industry.
Any one of the following:			
Dilute acetic acid (vinegar).	1	Salad dressing, developing photographic films.	Food stores, fermented apple cider or wine, photography supply.
Sulfuric acid, battery acid (oil of vitriol).	1	Storage batteries, material processing.	Motor vehicles, industrial plants.
Hydrochloric acid 1 (muriatic acid).	1	Petroleum wells, pickling and metal cleaning, industrial processes.	Hardware stores, industrial plants.

Air-tight container
Spoon or stick for stirring

Caution: **Sulfuric acid and hydrochloric acid will burn skin and ruin clothing. The fumes will irritate nose passages, lungs and eyes. Wash with large quantities of water upon contact.**

c. *Preparation.*
   (1) Commercial rubber latex may be used without further treatments before adding it to gasoline.
   (2) Natural rubber latex will coagulate (form lumps) as it comes from the plant. Strain off the thick part for use in flame fuel. If the rubber latex does not form lumps, add a small amount of acid to coagulate the latex and use the rubbery lump for gelling. It is best to air-dry the wet lumps before adding them to gasoline.
      (a) *Using commercial rubber latex.*
         1. Place the latex and the gasoline in the container to be used for the gelled gasoline and stir.

Caution: **Keep material away from open flames.**

         2. Add the vinegar (or other acid) to the liquid in the container and stir again until the gel forms. Store in an air-tight container until ready to use.

   Note. Use gelled gasoline as soon as possible because it becomes thinner on standing. If the gel is too thin, reduce the gasoline content (but not below 85% by volume).

         3. Natural rubber latex coagulates readily. If acids are not available, use one volume of acid salt (alum, sulfates and chlorides other than sodium and potassium). The formic acid content of crushed red ants will coagulate natural rubber latex.

(b) *Using natural rubber latex.*
80 parts by volume of gasoline.
20 parts by volume of coagulated or dried rubber.
Let the rubber lump soak in the gasoline in a closed container two or three days until a gelled mass is obtained. Prepare the gelled gasoline using the above formulation. This gelled gasoline should be used as soon as possible after it has thickened sufficiently.

d. *Application.* See paragraph 0303.1.

0303.7 WAX SYSTEMS

a. *Description.* See paragraph 0303.1.
b. *Material and Equipment.*

Ingredient	Parts by volume	Used for	Common source
Gasoline	80	Motor fuel	Gas stations or motor vehicles.
Any one of the following waxes:			
Ozocerite, mineral wax, fossil wax, ceresin wax.	20	Leather polish, sealing wax, candles crayons, waxed paper, textile sizing.	Natural deposits, general and department stores.
Beeswax	20	Furniture and floor waxes, artificial fruit and flowers, wax paper, candles.	Honeycomb from bees, general and department stores.

Ingredient	Part by volume	Used for	Common source
Bayberry wax, myrtle wax.	20	Candles, soaps, leather polish, medicine.	Natural from myrica berries, general, department, and drug stores.
Lye	0.5	Drain cleaner, making of soap.	Food and drug stores, soap factories.

Two air-tight containers
Spoon or stick for stirring

Caution: **Lye causes severe burns to eyes.**

Note. Lye is also known as caustic soda or sodium hydroxide. Allow for strength of lye. If only 50% (as in Drano), use twice the amount indicated above. Potassium hydroxide (caustic potash, potassa) may be used in place of lye.

c. *Preparation.*

(1) *Wax from natural sources.*

(a) Plants and berries are potential sources of natural waxes. Place the plants and/or berries in boiling water. The natural waxes will melt. Let the water cool, and the natural waxes will form a solid layer on the water surface. Skim off the wax and let it dry.

(b) Natural waxes which have suspended matter should be melted and screened through a cloth.

(2) *Gel from gasoline and wax.*

(a) Put the gasoline in a clean container.

Caution: **Keep material away from open flames.**

(b) Melt the wax and pour it into the gasoline container.

(c) Tightly cap the container and place it in hot water (sufficiently hot so that a small piece of wax will melt on the surface).

(d) When the wax has dissolved in the gasoline, place the capped container in a warm water bath and permit it to cool slowly to air temperature.

(e) If a solid paste of gel does not form, add another 10 parts by volume of melted wax and repeat (b), (c), and (d) above.

(f) Continue adding wax (up to 40 parts by volume) as before until a paste or gel is formed. If no paste forms at 80 parts by volume of gasoline and 40 parts by volume of melted wax, the wax is not satisfactory for gelled gasolines and may be used only in combination with alkali.

(3) *Gel from gasoline, wax and alkali.*
   70 parts by volume of gasoline
   29.5 parts by volume of melted wax
   0.5 parts by volume of staurated lye solution

(a) Prepare the saturated lye solution by *carefully* adding one volume of lye (or two volumes of Drano) to one volume of water and stir with a glass rod or wooden stick until the lye is dissolved.

Caution: **Lye causes severe burns to eyes. Add the lye to the water slowly. Let cool to room temperature and pour off the saturated liquid solution. Do not prepare this solution in an aluminum container.**

(b) Prepare the gasoline-wax solution according to the method described above.

(c) After the solution has cooled for about 15 minutes, CAUTIOUSLY loosen the cap, remove it and add the saturated lye solution.

(d) Stir about every five minutes until a gel forms. If the gel is not thick enough, remelt with another 5 parts by volume of wax and 0.1 part by volume of saturated lye solution. Stir contents as before. Store in an airtight container until ready to use.

Note. In addition to the listed waxes, the following may be used: candelilla wax, carnauba (Brazil) wax, Chinese (insect) wax, Japan (sumac) wax, montan (lignite) wax, and palm wax.

d. *Application.* See paragraph 0303.1.

0303.8 ANIMAL BLOOD SYSTEMS

a. *Description.* See paragraph 0303.1.

b. *Material and Equipment.*

Ingredient	Parts by volume	Chemical name	Used for	Common source
Gasoline	68		Motor fuel	Gas stations or motor vehicles.
Animal blood (sheep, cow, hog, dog, etc.).	30		Food, medicine.	Slaughter houses, natural habitat.
Any one of the following:				
Coffee (not decaffeinized).	2		Food, caffeine source, beverage.	Coffee processors, food stores.
Leaf tea	2		Food, beverage.	Tea processors, food stores.

74   TAGO 7189-C

Ingredient	Parts by volume	Chemical name	Used for	Common source
Lime	2	Calcium oxide.	Mortar, plaster, medicine, ceramics, industrial processes.	From calcium carbonate, hardware and drug stores.
Baking soda	2	Sodium bi-carbon-ate.	Baking, beverages, medicine, industrial processes.	Food and drug stores.
Epsom salts.	2	Magnesium sulfate hepta-hydrate.	Medicine, industrial processes, mineral water.	Natural deposits, drug and food stores.

Two air-tight containers
Spoon or stick for stirring

   c. Preparation.
      (1) *Animal blood serum.*
         (a) Slit animal's throat by jugular vein. Hang upside down to drain.
         (b) Place coagulated (lumpy) blood in a cloth or on a screen and catch the red fluid (serum) which drains through.
         (c) Store in a cool place if possible.
   Caution: **Animal blood can cause infections. Do not get aged animal blood or the serum into an open cut.**
      (2) *Preparation of gelled gasoline.*
         (a) Pour the animal bolod serum into a clean container and add the gasoline.
   Caution: **Keep material away from open flames.**

(b) Add the lime and stir the mixture for a few minutes until a firm gel forms. Store in an air-tight container until ready to use.

Note. Egg white may be substituted for up to 1/2 of the animal blood serum.

d. *Application.* See paragraph 0303.1.

0304. PARAFFIN-SAWDUST

a. *Description.*

(1) This item consists of a mixture of paraffin wax and sawdust. It is easily prepared and safe to carry. It is used to ignite wooden structures including heavy beams and timbers. It will also ignite paper, rags and other tinder type materials to build larger fires.

(2) This incendiary can be safely ignited by a match flame. However, any igniter listed in chapter 3 can be used in conjunction with specific delay mechanisms in chapter 5 for delayed ignition of this incendiary.

(3) All or part of the paraffin wax may be replaced by beeswax but *not* by vegetable or animal fats or greases.

b. *Material and Equipment.*

Paraffin wax, beeswax, or wax obtained by melting candles.
Sawdust.
Source of heat (stove, hot plate).
Pot.
Spoon or stick for stirring.

c. *Preparation.*

(1) Put enough wax in the pot so that it is about half full.

(2) Heat the pot on a stove or hot plate until the wax melts.

(3) Remove the heated pot from the stove or hot plate and shut off the source of heat. Add the sawdust to the melted wax until the pot is nearly full. Stir the mixture with a spoon or stick for a few minutes, being sure there is no layer of wax at the bottom of the pot which has not been mixed with the sawdust.

(4) While the mixture is in a fluid state, pour it into a waxed paper carton or other container. Upon cooling, the wax mixture will harden and take the shape of the container. The mixture can be stored for months without losing its effectiveness. If it becomes wet, it will be effective again when it is dried.

(5) A less effective incendiary may be made by melting some paraffin or beeswax, dipping sheets of paper in the molten wax for a few seconds, and removing the paper to let the wax harden. This waxed paper lights readily from a match. Although not as hot or persistent or the paraffin-sawdust mixture, the waxed paper is an excellent incendiary and may be substituted in many instances for initiating readily ignitable materials. The paper may be wadded up, folded, or torn into strips.

d. Application.

(1) Place about a quart of the mixture in a paper bag and put the bag down on the object to be burned. A match may be used to ignite the bag which will then ignite the paraffin-sawdust mixture. The fire starts very slowly so there

is no hazard involved, and it usually takes two or three minutes before the paraffin-sawdust mixture is burning strongly. This, of course, is a disadvantage where a hot fire is required quickly. Once started, however, this mixture burns vigorously because the paraffin itself gives a fairly hot flame and the sawdust acts like charcoal to increase the destructive effect.

(2) Where very large wooden beams or structures are to be burned use more of the mixture. A bag containing two or three quarts will be enough to set fire to almost any object on which paraffin-sawdust mixture can be used effectively.

(3) To be most effective on wood structures, this mixture should be in a pile, *never* spread out in a thin layer. If possible, place it under the object. When placing the incendiary in a packing box or in a room, place it in a corner.

0305. FIRE-BOTTLE (IMPACT IGNITION)

a. Description.

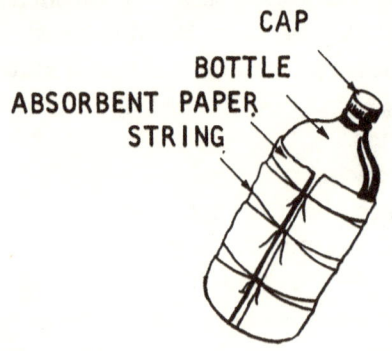

This item consists of a glass bottle containing gasoline and concentrated sulfuric acid. The exterior of the bottle is wrapped with a rag or absorbent paper. Just before use, the rag is soaked with a saturated solution of granulated sugar and potassium chlorate. Thrown against hard-surfaced targets such as tanks, automotive vehicles or railroad boxcars, this fire bottle is a very effective incendiary.

    b. Material and Equipment.
        Concentrated sulfuric acid (para 0103).
        Gasoline.
        Potassium chlorate (powdered).
        Sugar (granulated).
        Jar or bottle, with cap or stopper (½ pint).
        Cloth or absorbent paper.
        Jar or bottle, with cap or stopper (1 quart).
        String or tape.
        Heat resistant glass or porcelain pot (1 pint capacity).
        Heat source.
        Glass funnel.
        Spoon.
        Small container for measuring.
    c. Preparation.
        (1) Using the funnel, pour the gasoline into the quart bottle until approximately two-thirds full.

    Caution: **Keep this material away from open flames.**

        (2) Slowly add concentrated sulfuric acid through the funnel to the gasoline in the bottle and fill the bottle to within one inch of the top. The funnel must be used to direct the concentrated acid into the gasoline in the center

of the bottle. Stopper or cap the bottle securely.

Note. If only battery-grade sulfuric acid is available, it must be concentrated. See instructions under paragraph 0103.

(3) Flush the tightly capped bottle with water to remove any gasoline or acid adhering to the outside surface and dry the bottle. This *must* be done to avoid accidental combustion during the following steps.

(4) Wrap a clean cloth or several sheets of absorbent paper around the bottle. Fasten with strings or rubber bands.

(5) Prepare a saturated solution of granulated sugar and potassium chlorate in water as directed below.
(6) Add six measures of water to the porcelain pot and dry the measuring container with a clean rag or paper towel.
(7) Bring the water to a boil.
(8) Using a clean, dry spoon, place granulated sugar in the measuring container and add one and one-half measures of sugar to the boiling water.
(9) Wipe the spoon with a clean rag or paper towel and place one measure of potassium chlorate into the boiling sugar water.
(10) Remove the pot of boiling mixture immediately from the heat source and shut off heat source.
(11) When the solution is cool, pour it into the small ½ pint bottle using the glass funnel and cap tightly.
(12) Flush this bottle with water to remove any solution or crystals adhering to the outside surface and dry the bottle. When the crystals settle, there should be about ⅛ liquid above the crystals.

Caution: **Store this bottle separately from the other bottle containing gasoline and concentrated sulfuric acid.**

d. *Application.*
(1) Just prior to actual use, shake the bottle containing the sugar-potassium chlorate crystals and pour onto the cloth or paper wrapped around the gasoline-acid bottle. The fire bottle can be used while the cloth is still wet or

after it has dried. However when dry, the sugar-potassium chlorate mixture is very sensitive to sparks, open flame, bumping and scraping. In the dry condition the bottle should be handled carefully.

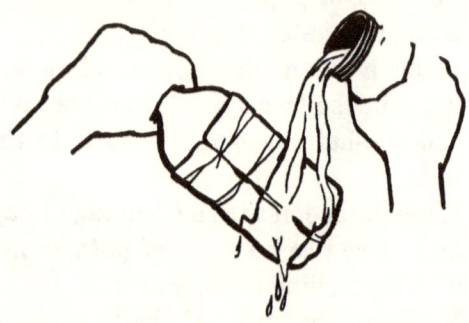

(2) The fire bottle should be gripped in one hand and thrown like a hand grenade. Upon impact with a metallic or other hard surface, the bottle will break and the sugar-potassium chlorate will react with the sulfuric acid. This reaction ignites the gasoline which will engulf the target area in flames.

0306. FIRE BOTTLE (DELAY IGNITION)

a. Description.

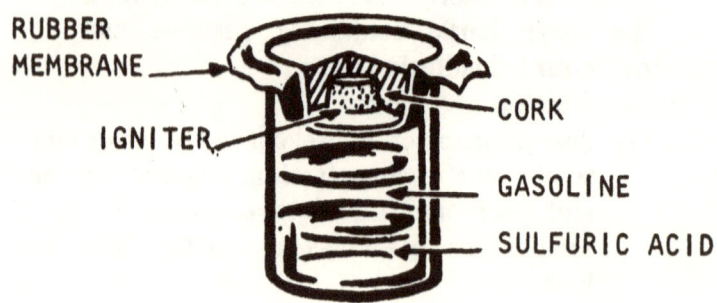

(1) This item consists of a bottle of gasoline and concentrated sulfuric acid which is ignited by the chemical reaction of the acid with Sugar-Chlorate Igniter (0201). A delay feature is incorporated in this incendiary. The amount of delay is determined by time it takes the sulfuric acid to corrode a rubber membrane and react with the igniter mix. Immediate ignition may also be achieved by breaking the bottle and allowing the ingredients to mix.

(2) Prepared fire bottles are stored upright. This allows the heavier acid to lay on the bottom, with the gasoline on top. When put in use, the bottle is inverted, allowing the acid to come in contact with the rubber membrane and to begin corroding it.

b. *Material and Equipment.*
Wide mouth bottle.
Cork or rubber stopper (must fit snugly in bottle).
Sheet rubber on rubber membrane.
Sugar-Chlorate Igniter (0201).
Concentrated Sulfuric Acid (0103).
Gasoline.

c. *Preparation.*
(1) Cut or drill a cavity on the bottom of the cork big enough to hold at least two teaspoonfuls of sugar-chlorate igniter. Be careful not to break through the cork. If the hole does go all the way through, it must be sealed with another smaller cork.

(2) Fill the bottle with a 50/50 concentration of gasoline and sulfuric acid. Pour the gasoline in first, then add the sulfuric acid carefully, making certain not to splash acid on the skin or in the eyes.

Note. If only battery grade sulfuric acid is available it must be concentrated before it can be used. See paragraph 0103 for details of concentration process.

(3) Fill the hole in the cork with Sugar-Chlorate Igniter (0201). Cover the side of the cork containing the igniter with a piece of thin rubber membrane and then force the cork into the gasoline-acid filled bottle. Take care to prevent any of the igniter mix from falling into the jar.

d. *Alternate Method of Preparation.*
    (1) Drill or cut a hole all the way through the cork.
    (2) Fill the bottle with gasoline and acid as described above.
    (3) Place the rubber membrane over the cork and install in the bottle. Make certain that cork is fitted tightly and rubber membrane fully covers the inner portion of the bottle.

(4) Fill the hole in the cork with igniter mixture as before and install a small cork in the hole covering the igniter mixture.

e. Application.
(1) To start the delay working invert the bottle. The acid will begin corroding the rubber membrane. When the acid breaks through, it will react violently and either break the bottle or blow out the cork stopper and ignite the gasoline.
(2) The Delay Fire Bottle works well on readily ignited materials where the scattering of the burning gasoline will start a number of fires at once. To ignite wooden structures, preparation such as piling up of flammable tinder and kindling is required.
(3) The delay time for initiation of the gasoline is slowed down in cold weather and may be stopped if the acid freezes. Check the delay time by testing the acid with the identical thickness rubber membrane at the temperature of expected use. Always use concentrated sulfuric acid.

0307. THERMITE

a. Description.
(1) Thermite is composed of magnetic iron flakes and aluminum powder. Thermite may be obtained as a manufactured item or may be improvised for use in welding machinery parts together and burning holes in metal structures. The termite reaction is initiated by strong heat and therefore cannot be directly ignited with a safety fuse or match.

The following igniters, found in chapter 3, may be used to initiate thermite: Powdered Aluminum—Sulfur Pellets (0207), Magnesium Powder—Barium Peroxide Igniter (0210), and Subigniter for Thermite (0211).

(2) Thermite is very safe to handle and transport because of its high ignition temperature. It burns well in cold and windy weather. Thermite will penetrate a sealed metal container and ignite the contents. It may be easily improvised if aluminum powder and iron oxide particles of the proper size are available.

b. *Material and Equipment.*

Aluminum powder (no coarser than ground coffee).
Iron oxide flakes (Fe_3O_4—similar to coarse ground coffee).
Spoon or cup for measuring.
Jar or can with tight fitting lid.
Cardboard can with metal ends.

c. *Preparation.*

(1) Place three parts by volume of iron oxide and two parts by volume of aluminum powder into the jar. Leave enough empty space to facilitiate mixing.

(2) Tighten the lid on the jar, turn the jar on its side and slowly roll until the two powders are completely mixed. The mixture is now ready for use and may be stored for months in the sealed container.

d. *Application.*

(1) Thermite is used to attack metallic targets such as transformers, electric motors, file cabinets, gears, bearings, boilers, storage tanks

and pipelines. In operation, the methods described below produce a quantity of molten metal that streams out the bottom of the unit. On contact with the target, the molten metal will cut through the outer metal casing and pour molten metal on the interior. Thermite is *not* recommended for use on moderate or heavy wooden structures or other applications where a persistent flame is required. Two basic techniques are described, one for burning holes in steel and the other for welding steel parts together.

(a) *Burning holes.*

1. In order to penetrate a steel plate with the minimum quantity of thermite, the mass of ignited thermite must be held away from the target during the initial combustion period. This minimizes conductive heat loss (from the thermite to the target) during this period and results in the thermite attaining maximum combustion temperature. When that temperature is reached, the thermite is dropped onto the steel plate surface and a hole is burned through the plate. The following illustrates the method for burning a hole through a plate of $3/8$ inch structural steel.

2. Cut a cardboard can (having metal ends) into two equal sections. Example of the type of cardboard container required are which contain household abrasive cleaners such as *AJAX*, *BON AMI* and *OLD DUTCH CLEANSER*.

3. One section of the can trimmed to a height of 2 inches and two side vents are cut as shown below.

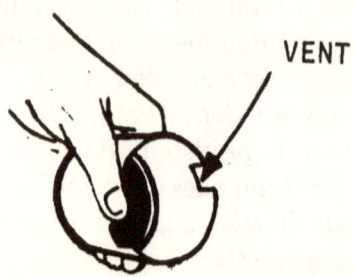

4. The other section is filled approximately ⅔ full with thermite. The thermite is then completely covered with one of the three above mentioned igniter materials to a depth of at least ¼ inch. Place the end of a length of Fuse Cord (0101) into the igniter mix, making certain that it does not extend into the thermite itself but ends in the center of the igniter mixture. Improvised String Fuse (0102) may be substituted for the Fuse Cord (0101) if desired.

5. The final assembly is constructed by placing the vented section, open face down, over the target area. The metallic end of this section is now facing up, away from the target surface. The section filled with thermite, igniter, and fuse is placed on top of the vented section. Both metal ends of the cardboard can are now touching.

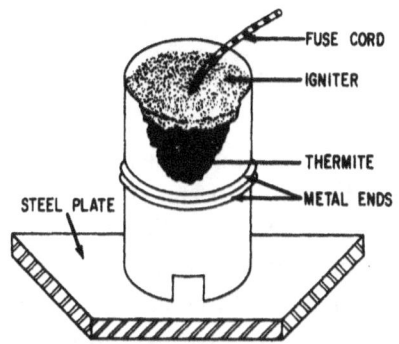

6. After ignition, the thermite burns a hole through the steel plate dropping extremely hot particles of molten slag into the interior of the steel container. The side vents cut in the bottom section of the can allow excess slag to run off and not close up the hole in the steel target.

(b) *Welding.*

1. A different method is employed when thermite is used to weld machinery components or plates together. The procedure is similar to that used for burning through steel except that the bottom stand-off is eliminated and the amount of thermite can be less than that used to burn through a ⅜ inch steel plate. The assembly is shown below.

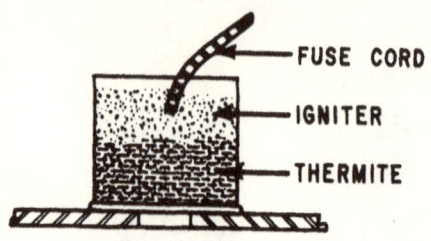

2. In this instance, heat is conducted from the thermite to the steel during the combustion period. Thus, the steel is heated to nearly the same temperature as the slag and a weld is effectively made.

Caution: **Never attempt to ignite thermite without at least a few seconds delay time because it burns so quickly and so hot that the user could be seriously burned if he were nearby when ignition took place.**

0308. FLAMMABLE LIQUIDS

a. Description. Flammable liquids are an excellent incendiary for starting fires with easily combustible material. They burn with a hot flame and have many uses as incendiaries. Most of these liquids are readily available and they are easily ignited with a match. However, these liquids tend to flow off the target and their characteristic odor may cast suspicion on the person found carrying them.

b. Material and Equipment.

Can or bottle with tight fitting lid (minimum 1 pint capacity).

One of the following volatile liquids:

Gasoline	Cleaners naptha
Kerosene	Turpentine

Toluene or Toluol   Lighter fluid
Xylene or Xylol   Fuel oil
Benzene or Benzol   Alcohol

c. Preparation. No preparation other than placing the liquid into an air-tight container for storage and transportation to the target is required.

d. Application. The most effective way to use flammable liquids is to pour at least a pint of the liquid on a pile of rags or sawdust which have been place in a corner of a packing box or other wooden object. This procedure keeps the liquid concentrated in a small area and gives a more persistent flame for igniting wooden objects. If small pieces of charcoal are available, they should be soaked with the flammable liquid and placed on the target. The charcoal will ignite and give a hot, persistent glow that is long lasting. After placing the flammable liquid on the target, throw a lighted match on a soaked area. Do not stand too close when this is done.

0309. INCENDIARY BRICK

  a. Description.

    (1) This incendiary is composed of potassium chlorate, sulfur, sugar, iron filings and wax. When properly made, it looks like an ordinary building brick and can be easily transported without detection. The incendiary brick will ignite wooden walls, floors, and many other combustible objects.

    (2) This incendiary can be directly ignited by all igniters listed in chapter 3, coupled with a specific delay mechanism found in chapter 5. To ignite this incendiary with White Phosphorus Solution (0209), the solution must first

be poured on absorbent paper and the paper placed on top of the brick.

b. *Material and Equipment.*

	Parts by volume
Potassium chlorate (powdered)	40
Sulfur (powdered)	15
Granulated sugar	20
Iron filings	10
Wax (beeswax or ordinary candle wax)	15

Spoon or stick
Brick mold
Red paint
Measuring cup or can
Double boiler
Heat source (hot plate or stove)

c. *Preparation.*

(1) Fill the bottom half of the double boiler with water and bring to a boil.

(2) Place the upper half of the boiler on the lower portion and add the wax, sulfur, granulated sugar, and iron filings in the proper amounts.

(3) Stir well to blend all the materials evenly.

(4) Remove the upper half of the double boiler from the lower portion and either shut off the heat source or move the upper section several feet from the fire.

Caution: **Extreme care should be exercised at this point because accidental ignition of the mixture is possible. Some means of extinguishing a fire should be at hand, a fire extinguisher or sand. It is important to keep face, hands, and clothing at a reasonably safe distance during the remainder of the preparation. A face shield and fireproof gloves are recommended.**

(5) CAREFULLY add the required amount of potassium chlorate and again stir well to obtain a homogeneous mixture.

(6) Pour the mixture into a brick mold and set aside until it cools and hardens.

(7) When hard, remove the incendiary from the mold, and paint it red to simulate a normal building brick.

d. *Application.*

(1) When painted, the incendiary brick can be carried with normal construction materials and placed in or on combustible materials.

(2) A short time delay in ignition can be obtained by combining Fuse Cord (0101) or Improvised String Fuse (0102) and one of the igniter mixtures found in chapter 3. (For example, several spoonfuls of Sugar-Chlorate mixture (0201) are placed on the incendiary brick. Fuse cord is buried in the center of the igniter mixture and the fuse is taped to the brick. Delay times are determined by the length of the fuse. Suitable delay mechanisms are given in chapter 5 for delay times longer than those practical with fuse cord.)

CHAPTER 5
DELAY MECHANISMS

0401. CIGARETTE

a. Description.

(1) This item consists of a bundle of matches wrapped around a lighted cigarette. It is placed directly on easily ignited material. Ignition occurs when the lighted portion of the burning cigarette reaches the match heads. This delay mechanism can be used to initiate all igniters listed in chapter 3 except Magnesium Powder—Barium Peroxide (0210) and Powdered Aluminum—Sulfur Pellets (0207). A cigarette delay directly ignites the following incendiaries: Napalm (0301), Gelled Gasoline (exotic thickeners) (0302), and Gelled Gasoline (improvised thickeners) (0303).

(2) The following *dry* tinder type materials may also be directly ignited by the cigarette delay mechanism: Straw, paper, hay, woodshavings and rags.

(3) Usually this delay will ignite in 15 to 20 minutes, depending on length of cigarette, make of cigarette, and force of air currents. A duplicate delay mechanism should be tested to determine delay time for various ambient conditions.

(4) The cigarette must be placed so that the flame will travel horizontally or upward. A burning cigarette that is clamped or held will not burn past the point of confinement. Therefore, the cigarette should not contact any object other than matches.

b. *Material and Equipment.*
Cigarette.
Matches (wooden).
Match box.
String or tape.

c. *Preparation.*
(1) *Picket-fence delay.*

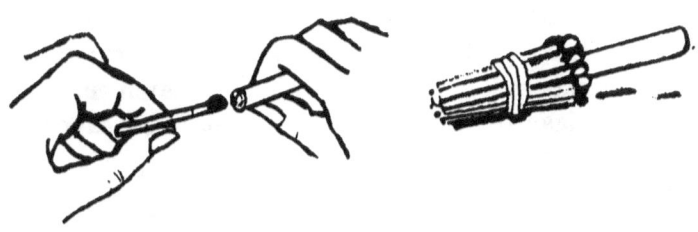

(a) Push one wooden match head into a cigarette a predetermined distance to obtain the approximate delay time.

(b) Tie or tape matches around the cigarette with the match heads at the same location as the first match in the cigarette.

(2) *Match box delay.*

Tear out one end of the inner tray of a box of matches (the end next to the match heads). Push one match into the cigarette. Insert this cigarette into the bunch of matches and parallel to the matches at the center of the pack. Slide the tray out of the inner box, leaving the match heads and the cigarette exposed. The head of the match in the cigarette should be even with the exposed match heads.

d. Application.

(1) *Picket-fence delay.*

(a) Light the cigarette and place the delay mechanism on a pile of igniter mixture, paper, straw, or other dry tinder type material. Be sure that the portion of the cigarette between the lit end and the match heads is not touching anything.
(b) Pile tinder material all around the cigarette to enhance ignition when the match heads ignite.

(2) *Match box delay.*

(a) Place the delay so that the cigarette is horizontal and on top of the material to be ignited. Light the cigarette.
(b) Be sure ignitable material such as paper, straw, flammable solvents, or napalm is placed close to the match heads. When using flammable solvents, light the cigarette away from the area of solvent fumes.
(c) To assure ignition of the target, sprinkle some igniter material on the combustible material. The match box delay is then placed on top of the igniter material.

0402. GELATIN CAPSULE

a. Description.

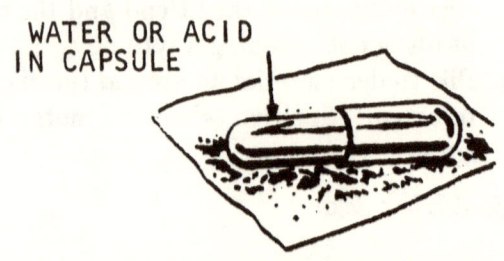

Gelatin capsule delays work by the action of either water or concentrated sulfuric acid on the gelatin. When the liquid dissolves the gelatin, it contacts and reacts with an igniter mix. These delays can be used with various igniters, are easily prepared and easily carried. The disadvantage is that the delay times vary with temperature and they will not work at or below 32° F. Gelatin capsule delays will work with the following igniters:
    (1) Water actuated igniters such as Sugar—Sodium Peroxide (0203), Silver Nitrate—Magnesium Powder (0208), and Aluminum Powder—Sodium Peroxide (0204).
    (2) Concentrated sulfuric acid actuated igniters such as Sugar-Chlorate (0201), Fire Fudge (0202), Sugar—Sodium Peroxide (0203) Aluminum Powder—Sodium Peroxide (0204), Match Head (0205), and Silver Nitrate—Magnesium Powder (0208).

b. Material and Equipment.
    Concentrated sulfuric acid or water.
    Gelatin capsules (1 fluid ounce capacity).

Igniter mixture.
Glass jar or bottle with glass or plastic stopper for carrying acid.

c. *Preparation.*
   (1) Fill the gelatin capsule with either water or sulfuric acid, depending on which igniter is being used. Use a medicine dropper to fill the capsule. Wipe the outside of the capsule carefully and place it on a quantity of igniter mixture.
   (2) Gelatin will slowly dissolve in either water or concentrated sulfuric acid, usually faster in water than in acid. Sulfuric acid should be handled carefully and only in glass or unchipped enamel containers.

d. *Application.*
   (1) Fill a gelatin capsule with one of the igniter mixes listed under *Description* above. Once the liquid is added to the capsule, the next operations should be done quickly. Pile the igniter mixture on and around the capsule. Then place incendiary material in contact with the igniter mixture. (In damp weather this method should not be used with water activated igniters because premature ignition may be caused by humidity in the air.)
   (2) Use the following method in damp weather. Fill a gelatin capsule with one of the igniter mixes listed above. Be sure that both halves of the capsule fit tightly and that no igniter mix is clinging to the outside of the capsule. Place the capsule in a shallow glass or porcelain dish filled with water or concentrated sulfuric acid, depending on which type of

igniter mix is used. Make sure the capsule is touching one edge of the bowl and quickly pile incendiary material close to the capsule so that when the capsule ignites, the incendiary will also ignite.

(3) The gelatin capsule delays work slowly in cold weather and will not work at or below 32° F. Capsule thickness also affects delay time. In water at 77° F., a delay time of approximately 20 minutes can be expected, while the same type of capsule in concentrated sulfuric acid at 77° F. will give a delay time of approximately one hour. At a temperature of 50° F., the same type of capsule will give a 6 to 8 hour delay time in water and about 24 hours delay time in concentrated sulfuric acid. Delay times become less accurate at lower temperatures.

(4) The above listed delay times are given for one type of gelatin capsule only. Various types of capsules will give various delay times. Therefore, always check delay times for the capsule to be used.

(5) The sulfuric acid must be concentrated. If only battery-grade sulfuric acid is available, it must be concentrated before use to a specific gravity of 1.835 by heating it in an enameled, heat resistant glass or porcelain pot until dense, white fumes appear. See paragraph 0103 for details.

0403. RUBBER DIAPHRAGM

a. *Description.*

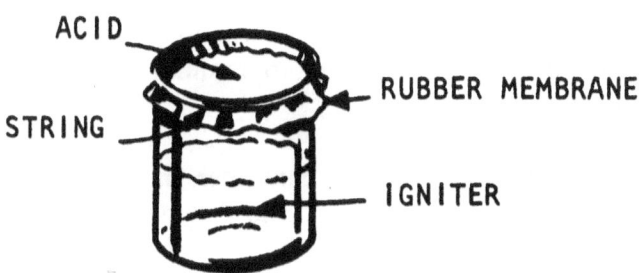

(1) This delay operates by the action of concentrated sulfuric acid on a thin rubber diaphragm. As the acid eats through the diaphragm, it drips onto the igniter mix and combustion results. This delay can be used to initiate the following igniters listed in chapter 3: Sugar-Chlorate (0201), Fire Fudge (0202), Sugar—Sodium Peroxide (0203), Aluminum Powder—Sodium Peroxide (0204), Match Head (0205), Silver Nitrate—Magnesium Powder (0208), and Fire Bottle (0306).

(2) The delay does not burn or glow, a very desirable feature where premature detection may occur. The main disadvantages of this type of delay are—

  (a) Delay time fluctuates with temperature changes.

  (b) Delay is not reliable below 40° F.

  (c) Sulfuric acid involves hazards to the operator.

b. *Material and Equipment.*
   Concentrated sulfuric acid.
   Thin rubber (such as balloons or condoms).
   String, tape, or rubber bands.
   Glass jar with glass stopper for carrying acid.
   Wide-mouthed jar or can (approximately 1 pint capacity).
c. *Preparation.*
   (1) Fill the wide mouth container three-quarter full with any one of the following igniter materials:
   Sugar-Chlorate (0201).
   Fire Fudge (0202).
   Sugar—Sodium Peroxide (0203).
   Aluminum Powder—Sodium Peroxide (0204).
   Match Head (0205).
   Silver Nitrate—Magnesium Powder (0208).
   (2) Place the rubber diaphragm over the open end of the container and leave it loose enough to sag slightly into the jar. Either tie in place or secure with a rubber band.
   (3) Pour about 1 fluid ounce of concentrated sulfuric acid into a small glass jar with a glass stopper and seal tightly.
d. *Application.*
   (1) Place the jar with the rubber membrane at the desired target. Pile the material to be ignited around this jar so that when the flames issue from the jar, they will ignite the incendiary materials. *Do not put any of this igniter material on the rubber membrane.* Pour the 1 fluid ounce of concentrated sulfuric acid onto the rubber membrane. When the acid penetrates the rubber and drips onto the

igniter mix, a chemical reaction occurs and combustion results.

(2) The time delay of this device depends on the kind and thickness of rubber used, and on the ambient temperature. Test a similar device before actual use on the target.

(3) Using a thin rubber membrane such as a condom at a temperature of 77° F., a delay time of 15 to 20 minutes is normal. This same delay when tested at 40° F. may take as long as eight hours to penetrate the rubber membrane. Do not use this delay at temperatures below 40° F.

(4) Another simple method of using this type of delay is to first fill a small jar half full of concentrated sulfuric acid. Tie or tape a rubber membrane over the open end of the jar. BE SURE NO ACID CAN LEAK OUT. Place the bottle on its side, on top of a small pile of igniter material which will ignite on contact with the acid. When the acid penetrates the membrane, combustion will occur as before. If thicker rubber is used, stretch the rubber tightly over the mouth of the jar. This will decrease the delay time because the acid will attack the stretched rubber more effectively.

(5) A rubber glove may also be used as a membrane for this delay. Pour some concentrated sulfuric acid into the glove and suspend the glove over a pile of igniter material. When the acid eats through the glove, it will drip onto the igniter and start a fire. A rubber glove will give a longer delay time than a condom because the material is thicker.

IGNITER MIX

(6) The rubber membranes for use in this delay must be without pin holes or other imperfections. The sulfuric acid must be *concentrated*. If only battery-grade sulfuric acid is available, it must be concentrated before use to a specific gravity of 1.835 by heating it in an enameled, heat-resistant glass or porcelain pot until dense, white fumes appear. See paragraph 0103 for details.

0404. PAPER DIAPHRAGM (SULFURIC ACID)

a. Description.

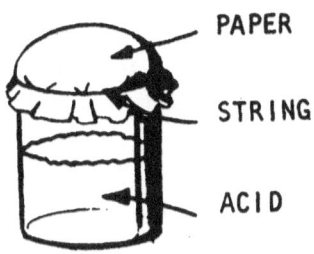

This device consists of a half-full jar of concentrated sulfuric acid, and a paper diaphragm. The paper diaphragm is a piece of paper tied securely over the mouth of the jar. When the jar is placed on its side, the acid soaks through or corrodes the paper. The acid then contacts the igniter material and causes it to burts into flames. This delay can be used for initiating the following igniters listed in chapter 3: Sugar-Chlorate (0201), Fire Fudge (0202), Sugar—Sodium Peroxide (0203), Aluminum Powder—Sodium Peroxide (0204), Match Head (0205), Silver Nitrate—Magnesium Powder (0208).

b. Material and Equipment
Wide-mouthed jar.
Sulfuric acid (concentrated).
Paper.
String.

c. Preparation. Remove the cap from a wide-mouthed jar Fill about half-full with concentrated sulfuric acid. Tie the paper securely over the mouth of the jar.

d. Application.

   (1) Make a pile of dry flammable material such as rags, papers, empty boxes, or cartons. Spread out a piece of absorbent paper on this material. Spread igniter material on the absorbent paper and place the jar (on its side) on top of the igniter material. Make certain the jar does not leak. When the acid soaks through or corrodes the paper, it will contact the igniter material and cause it to burst into flame.

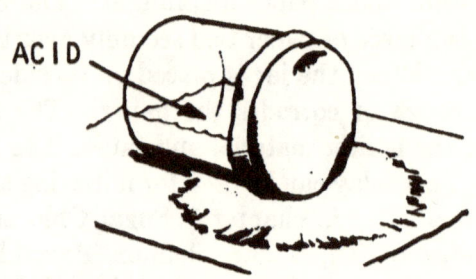

   (2) This device is not reliable at temperatures below 40° F. The time delay depends on the thickness of the paper. A similar device should be tested to determine the delay provided by various thicknesses of paper. It should be tested at the temperature at which it will be used, to be sure of positive ignition. Ignition should occur in about 2 minutes at 68° F. when using writing paper. Higher ambient temperatures shorten delay times, and lower temperatures lengthen delay times.

0405. PAPER DIAPHRAGM (GLYCERIN)

a. *Description.*
(1) This device consists of potassium permanganate crystals wrapped in layers of absorbent paper. Glycerin is brought into contact with the wrapped potassium permanganate crystals by slowly soaking through the paper. This wets the wrapped crystals causing combustion. This delay can be used for directly initiating all igniters listed in chapter 3 except White Phosphorus (0209). The igniting ability of this delay is increased when magnesium or aluminum particles are mixed with the potassium permanganate crystals.

(2) The following incendiaries (ch 4) can be directly ignited using this delay: Napalm (0301), Gelled Gasoline (exotic thickeners) (0302), Gelled Gasoline (Improvised thickeners) (0303), Paraffin-Sawdust (0304), and Incendiary Brick (0309). Other combustible dry materials such as paper, rags, straw, and excelsior can also be directly initiated. This delay is not recommended for use in tempertures below 50° F.

b. *Material and Equipment.*
Absorbent paper (toilet paper, paper, toweling, newspaper).
Glycerin (commercial grade).
Magnesium or aluminum particles (consistency of granulated sugar).
Rubber bands or string.
Small shallow dish.

Potassium permanganate (consistency of coarse ground coffee).
Small bottle (approximately 1½ fluid ounces).
Spoon (perferably nonmetallic).

c. Preparation.
  (1) Fill the small bottle with glycerin.
  (2) Wrap a quantity of potassium permanganate crystals (a mixture of 85 parts potassium permanganate and 15 parts magnesium or aluminum particles can be substituted to produce a hotter flame) in absorbent paper. Make certain that none of the crystals fall out.
  (3) The bottle and package may be carried by the person without hazard to himself, and will be available for use when needed.

d. Application.
  (1) To use this delay, pour the glycerin into a small shallow dish or pan. Pile incendiary material around the dish so that when the glycerin ignites it will ignite the incendiary material. Place the paper container of potassium permanganate crystals into the pan of glycerin. When the glycerin soaks through the paper and contacts the potassium permanganate, ignition occurs within a few seconds.

GLYCERIN
BAG OF CRYSTALS

(2) By using various kinds of paper, different delay times can be obtained. Using more layers of paper for wrapping will increase the delay time. Using this delay at higher temperatures will also decrease the delay time. Delay times from one minute to approximately one hour are possible, depending on the conditions.

(3) The delay time should be checked under conditions which are similar to those expected at the target.

0406. CANDLE

  a. Description.

This delay ignites flammable fuels of low volatility such as fuel oil and kerosene. A lighted candle properly inserted in a small container of flammable liquid of low volatility causes ignition of the flammable liquid when the flame burns down to the liquid level. The flame from the burning liquid is used to ignite incendiary material such as paper, straw, rags, and wooden structures. The delay time is reasonably accurate, and may be easily calibrated by determining the burning rate of the candle. No special skills are required to use this delay. Shielding is required for the candle when used in an area of strong winds or drafts. This delay is *not* recommended for use with *highly volatile liquids* because premature ignition may take place. This device is useful where a delay of one hour or longer is desired. The candle delay works well in cold or hot weather, and has the advantage of being consumed in the resulting fire, thus reducing evidence of arson.

 b. Material and Equipment.
 Candle.
 Bowl.
 Perforated can or carton.
 Fuel oil or kerosene.
 Matches.
 Small piece of cloth.

 c. Preparation.
  (1) Make two marks on the side of the candle, 1½ inches and 2 inches from the top. Light the candle and record the times at which the wax melts at the marks on the side.

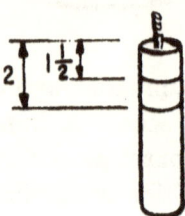

  (2) The distance burned by the candle divided by the elapsed time determines the burning rate of the candle.

d. Application.
   (1) Using a lighted candle of desired length, drip hot wax in the center of the bowl. Melt the base of the candle with a lighted match. Firmly press the softened base of the candle into the hot wax in the center of the bowl. Be sure the candle will stand up securely without toppling over. Extinguish the candle. Wrap a small piece of cloth around the candle and slide it down to the bottom of the bowl. Place a quantity of fuel oil or kerosene in the bowl. Be sure that the level of the fluid reaches the cloth, so it will act as a wick. Pile the incendiary material around the bowl where it can catch fire after the fuel oil or kerosene ignites.
   (2) If this delay must be set in a windy or drafty location, place a shield over it. Notch or punch holes in a metal can or cardboard carton at the bottom and sides for ventilation, and place this cover over the delay.

0407. OVERFLOW
   a. Description.

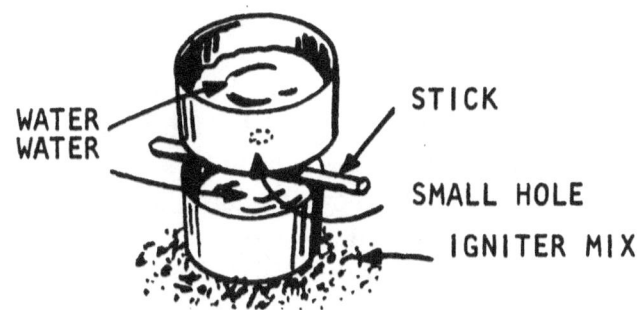

This item provides a time delay in starting a fire. It consists of two tin cans, with tops removed, and uses either water or glycerin to activate the igniter material. A hole is punched in the closed end of one can. This can is placed on top of the other can which is partially filled with the liquid. The top can is completely filled with the liquid. When the bottom can fills and overflows, the overflowed liquid will react with the igniter material placed around the bottom can. This device is used for igniting the following water actuated igniters listed in chapter 3: Sugar—Sodium Peroxide (0203), Aluminum Powder—Sodium Peroxide (0204), and Silver Nitrate—Magnesium Powder (0208). Glycerin is used as the initiating liquid to ignite Potassium Permanganate glycerin (0206).

   b. Material and Equipment.
      Two tin cans.
      Nail or punch.
      Hammer.
      Water or glycerin.
      Can opener.
   c. Preparation.
     (1) Remove the tops from two cans.
     (2) Punch or drill a small hole in the closed end of one of the cans.
     (3) Partially fill the other can with either water or glycerin.
     (4) Place the can with the hole in the bottom on top of the can partially filled with igniting fluid. Insert a twig or small stick between the two cans to allow the liquid to overflow from the bottom can.
     (5) Fill the upper can with the same igniting fluid as that previously placed in the bottom can

and determine the time required for the fluid to overflow from the bottom can. If two cans of the same size are used, either one may be used for the top. If different size cans are used, place the larger can on top. The delay is variable and adjustable depending on the sizes of the cans, the quantity of liquid used, or the diameter of the hole in the top can.

 d. *Application.*

 (1) Always test the glycerin delay at the temperature at which it will be used. Glycerin flows slowly when cold. Do not use water in this delay near or below its freezing point, 32° F.

 (2) Place the delay in the target area and fill both upper and lower cans to the desired level with the appropriate liquid.

 (3) Pile igniter material around the bottom of the overflow can so the activating liquid can easily make contact with the igniter material as it flows down the side of the can.

0408. TIPPING DELAY—FILLED TUBE

 a. *Description.*

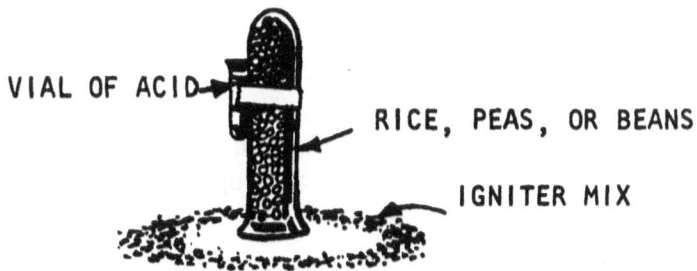

(1) This delay is composed of a hollow metal rod or bamboo filled with wet beans, rice or peas. The tube is inverted and placed in the center of a ring of igniter material and a small vial of water or acid is tied to the tube. When the wet beans expand, they lift and topple the tube, thereby spilling the acid or water onto the igniter causing combustion.

(2) This tipping delay may be used with a variety of igniters. They are easily prepared, and give fairly accurate delay times. This delay should not be used at temperatures near or below 32° F. when water is used as the initiator due to freezing. The following water actuated igniters listed in chapter 3 can be used with this mechanism: Sugar—Sodium Peroxide (0203), Aluminum Powder—Sodium Peroxide (0204) and Silver Nitrate—Magnesium Powder (0208). The delay may be used with concentrated sulfuric acid to initiate the above igniters and the following acid activated igniters: Sugar-Chlorate (0201), Fire Fudge (0202), and Match Head (0205). This delay may be used with the Glycerin—Potassium Permanganate Igniter (0206).

b. *Material and Equipment.*

Metal tube, pipe or piece of bamboo closed at one end, 4 to 6 inches long and 1 inch inside diameter, or glass test tube of similar dimensions.

Small glass vial or bottle with open mouth of 1 fluid ounce capacity.

String or rubber bands.

Rice, peas, or beans.

Water.

Concentrated sulfuric acid.

c. Preparation. The pipe or tube may be made of any material. It must be closed at one end and flat at the other in order to stand vertically. A large glass test tube is ideal for this purpose.

   (1) Using some string or rubber bands, attach the small vial to the larger tube. Attach the vial near the top with the open end of the vial pointing up and the open end of the tube down.

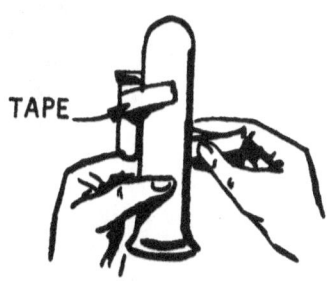

   (2) This assembly should stand up without toppling over. If it appears unsteady, move the vial downward slightly. A final adjustment may be required when the delay is filled with the required materials.

d. Application.

   (1) Rice will usually give delays of about ten to twenty minutes. Peas and beans will usually give delay times up to 4 or 5 hours. Whichever is used it must be first tested to determine the delay time for the tube that will be used.

   (2) To use this device, tightly pack the piece of pipe or bamboo with rice, peas or beans

depending on what delay time is required. Add enough water to completely moisten the beans and quickly pour off the excess water. Place the pipe open end down, and immediately fill the small vial with water or concentrated sulfuric acid, depending on which igniter is being used.

(3) Place a quantity of the igniter mixture in a ring around the delay assembly. Make the ring of such diameter that when the tube falls over, the acid or water from the vial will spill onto the igniter mixture.

(4) Place incendiary material where the flame from the igniter will start it burning.

(5) Another way in which the tipping delay can be used is to fill the small vial with glycerin instead of water or acid and then spread potassium permanganate crystals in a ring around the delay. When the glycerin is spilled onto the crystals, combustion will occur and ignite the incendiary material. The glycerin igniter will not work in temperatures below 50° F.

(6) It is recommended that this device be tested at the same temperature at which it is to be used.

0409. TIPPING DELAY—CORROSIVE OR DISSOLVING ACTION

a. Description.

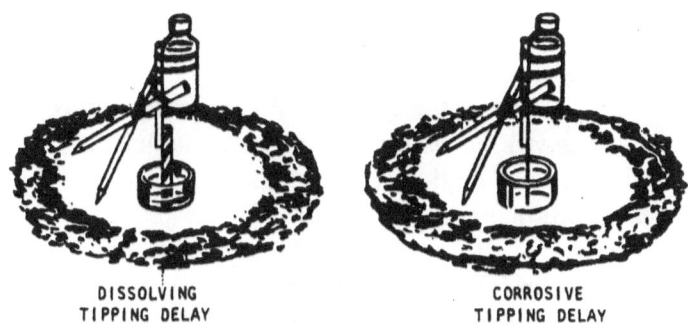

DISSOLVING TIPPING DELAY

CORROSIVE TIPPING DELAY

(1) This device consists of a vial of initiating liquid supported by a tripod. One of the legs which supports the vial of liquid is dissolved by a fluid. The center of gravity of the structure changes and the structure topples over. The contents of the vial spill onto an appropriate igniter mixture and combustion occurs.

(2) This corrosive or dissolving tipping delay may be used with a variety of igniters. However, it should not be used at temperatures near or below 32° F. when water is used as the initiator due to freezing of the water.

(3) The following water actuated igniters listed in chapter 3 can be used with this mechanism:

Sugar—Sodium Peroxide (0203), Aluminum Powder—Sodium Peroxide (0204) and Silver Nitrate—Magnesium Powder (0208). The delay may be used with concentrated sulfuric acid to initiate the above igniters and the following acid activated igniters: Sugar-Chlorate (0201), Fire Fudge (0202), and Match Head (0205). This delay may be used with the Glycerin—Potassium Permanganate Igniter (0206)'

b. *Material and Equipment.*

Three wooden sticks or wooden pencils (approximately 6 inches long by $\frac{1}{4}$ inch diameter).

Glass vial (1 fluid ounce capacity).

String, tape or rubber bands.

Any one of the igniter mixtures mentioned above.

One of the following combination of items:

(1) Long sticks of hard candy and water.

(2) Lengths of bare copper wire and concentrated nitric acid.

(3) Iron nails or wire approximately $\frac{1}{32}$ inch diameter by 4 inches long and concentrated hydrochloric acid.

(4) Iron nails or wire and saturated cupric chloride solution.

2 glass containers with glass stoppers for carrying acid.

Shallow glass or porcelain bowl such as soup bowl or ink bottle.

c. *Preparation.*

(1) Make a tripod out of three sticks, taping them together at the top. Two legs should be the same length; the third should be about 2—3 inches shorter.

(2) Tape to the short leg, either a stick of hard candy, piece of heavy bare copper wire, steel nail, or steel wire, adjusting the length so that the wire leg stands almost vertically.

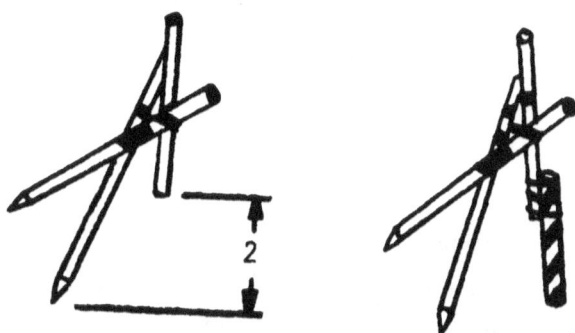

(3) The finished tripod should have a distance of about 4—5 inches between any two legs.
(4) To the top of the tripod, on the short leg, firmly tape or tie the small 1-fluide ounce capacity vial, open end up. Make certain that the tripod still stands upright after attaching the vial. The distance between legs may have to be varied to keep the tripod barely standing upright.

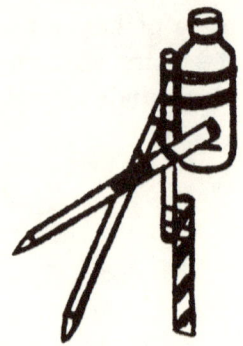

d. *Application.*

(1) To use the delay device, insert the leg of the tripod which has the candy, wire, or nails into a glass or porcelain bowl. Fill the vial at the top of the tripod with either water, concentrated sulfuric acid, or glycerin, depending on which igniter is being used. Spread a quantity of the proper igniter material in a ring around the tripod, placing it where the spilled initiating liquid is certain to contact it. Fill the glass or porcelain bowl with the prescribed liquid for dissolving the leg of the tripod in the bowl. For hard candy the liquid is water; for copper wire the liquid is concentrated nitric acid; for steel nails the liquid may be either concentrated hydrochloric acid, or a saturated solution of cupric chloride.

(2) No definite delay times can be established for these delays because of factors such as temperature, solution concentration, and imperfections in the leg of the tripod. Prior to use, test the device under conditions expected

at the target. The following table should be used merely as a guideline of expected delay times for the various materials.

Delay material	Delay time
Hard candy plus water	5—10 minutes
Copper wire plus concentrated nitric acid	2—5 minutes
Copper wire plus nitric acid diluted with an equal volume of water.	45—60 minutes
Steel wire or nails plus concentrated hydrochloric acid.	24 hours to 7 days
Steel wire or nails plus cupric chloride solution.	10 minutes to 5—6 hours.

   (3) The delay time will vary greatly with only moderate changes in temperature. Do not use this type of delay mechanism where accurate delay times are required.

0410. BALANCING STICK

a. Description.

   (1) This delay device consists of a piece of wood or stick, a small vial, a nail, a piece of string, and a long strip of cloth. A hole is drilled through the middle of the stick. The vial is

fastened to one end, and the strip of cloth to the other. The length of the cloth is adjusted so that the rod just balances on a nail passing through the hole when the vial is $3/4$ full. The cloth is wetted with solvent to make it heavy and the vial is filled with initiating liquid to maintain balance. As the solvent evaporates, the end of the stick which supports the vial of initiating liquid becomes heavier than the end supporting the cloth. The unbalanced stick rotates about the nail until the initiating liquid spills onto the igniter mixture and combustion occurs. Fire then spreads to and ignites incendiary material.

(2) This device may be used with a variety of igniters. However, it should not be used at temperatures near or below 32° F. when water is used as the initiator due to freezing of the water. The following water actuated igniters listed in chapter 3 can be used with this mechanism: Sugar—Sodium Peroxide (0203), Aluminum Powder—Sodium Peroxide (0204) and Silver Nitrate—Magnesium Powder (0208). The device may be used with concentrated sulfuric acid to initiate the above igniters and the following acid activated igniters: Sugar-Chlorate (0201), Fire Fudge (0202), and Match Head (0205). It may also be used with the Glycerin—Potassium Permanganate Igniter (0206).

b. *Material and Equipment.*
Piece of wood ($1/8$ by $1/8$ by 16 inches).
2 Nails.

String.
Strip of cloth.
2 glass vials (1 fluid ounce) with stoppers.
c. *Preparation.*
  (1) Drill a hole through the middle of the stick as shown below.

  (2) Insert a nail through the hole. The nail should permit the stick to turn freely. Tie a piece of string (4–6 inches in length) to both ends of the nail, forming a loop. It is not important that the stick balance exactly.
  (3) To one end of the stick tape a small glass vial. Tilt the vial when attaching it so that when this end of the stick is about 8 inches above the other end, the vial will be vertically upright. On the other end of the stick tie a strip of cloth, rag, or rope. This strip should be heavy enough so that the stick is balanced when the vial is about ¾ full of initiating fluid.

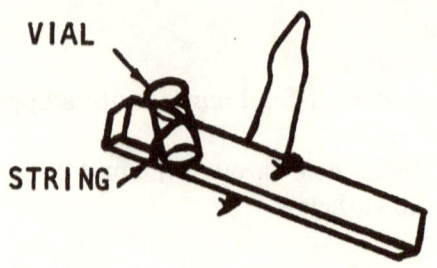

d. *Application.*
  (1) To use this delay, drive a nail (approximately 4 inches long) into a wall or wooden box about 8 inches above the floor, leaving at least 2 inches of the nail projecting. Place the loop of string on the nail near the head of the nail. The stick should not touch the box or wall, but must swing freely. The rag should touch the floor. Pour enough solvent on the rag to soak it thoroughly (approximately 1 fluid ounce). Working quickly, fill the vial with initiating liquid and balance the rod by shifting the cloth. Spread a quantity of appropriate igniter mixture on the floor where the initiating liquid will spill when the solvent on the cloth evaporates. In a few minutes the solvent will evaporate, causing the stick to become unbalanced. The vial will tilt with the stick and, the liquid in the vial will pour out and initiate the igniter mixture.

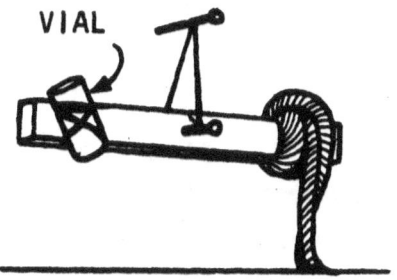

(2) Where no solvent is available or where the odor of solvent may make the device easy to detect, do not use cloth soaked with solvent. Use a wire basket containing ice as shown below.

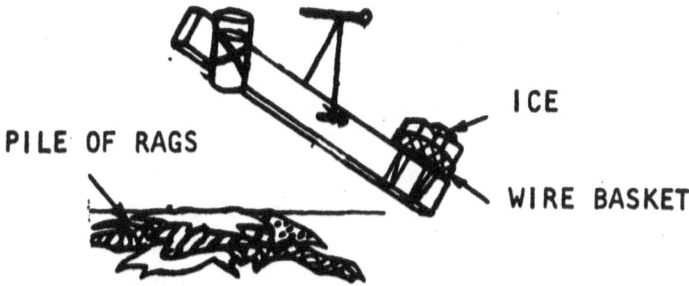

(3) When ice is used, the delay time will be a matter of minutes, depending on the ambient temperature. Ice cannot be used at temperatures near 32° F. Be sure that the drippings from the melting ice does not wet the igniter or interfere with initial combustion of flammable material.

0411. STRETCHED RUBBER BAND

a. Description.

This item utilizes a rubber band, which has been soaked in gasoline or carbon disulfide until it has considerably expanded. After removal of the rubber band from the solvent, the rubber band is attached to a wall and to a bottle containing igniter fluid. As the rubber band contracts due to solvent evaporation, the bottle is tipped and initiator liquid comes in contact with an appropriate igniter material. This stretched rubber band delay may be used with a variety of igniters. However, it should not be used at temperatures near or below 32° F. when water is used as the initiator because the water freezes. The following water actuated igniters listed in chapter 3 can be used with this mechanism: Sugar—Sodium Peroxide (0203), Aluminum Powder—Sodium Peroxide (0204), and Silver Nitrate—Magnesium Powder (0208). The delay may be used with concentrated sulfuric acid to initiate the above igniters and the following acid activated igniters: Sugar-Chlorate (0201), Fire Fudge (0202), and Match Head (0205). This delay may be used with Glycerin—Potassium Permanganate Igniter (0206).

b. *Material and Equipment.*
   Bottle or jar (1 to 2 fluid ounce capacity).
   Rubber bands.
   Gasoline or carbon disulfide.
   Air tight container for carrying the gasoline or carbon disulfide.
   Nails.
   Igniter.
c. *Preparation.*
   (1) Fill a bottle (1 to 2 fluid ounce capacity) with water, acid, or glycerin, depending on which igniter is to be used.
   (2) Soak the rubber bands in gasoline or carbon disulfide for about one hour. Do not soak too long or they will become excessively weakened.
d. *Application.*
   (1) At the place where the delay is to be used, drive a large headed nail into the wall, leaving about 2 to 2½ inches exposed. Loop the rubber bands over the head of the nail. Place the bottle two bottle heights away from the nail. Quickly loop the free end of the rubber bands over the neck of the bottle. Move the bottle back and forth until there is just enough tension in the rubber bands to hold the bottle without it toppling when a pencil or twig is placed under the far end. The stick under the end of the bottle is used as a tilt device to make sure that the bottle topples over when the rubber band contracts.
   (2) Place some incendiary material close to the bottle. Sprinkle a quantity of igniter mixture about the area in which the liquid will be spilled. As the solvent evaporates, the rubber

bands will shrink, tip the bottle, spill the liquid, and initiate the igniter material.

Note. Always set up the bottle before spreading the igniter mixture.

0412. ALARM CLOCK

a. Description.

(1) This device is used for igniting materials after a definite delay time. The device employs a manually-wound alarm clock, with the alarm bell removed, as the timing mechanism. A piece of string is fastened to the key used to wind the alarm. The other end of the string is fastened to a bottle of appropriate initiating liquid. When the modified alarm mechanism is tripped, the winding key will reel in the string and overturn the bottle of initiating liquid and start a fire.

(2) This alarm clock delay may be used with a variety of igniters. However, it should not be used at temperatures near or below 32° F. when water is used as the initiator because the water freezes. The following water actuated

igniters listed in chapter 3 can be used with this mechanism: Sugar—Sodium Peroxide (0203), Aluminum Powder—Sodium Peroxide (0204), and Silver Nitrate—Magnesium Powder (0208). The delay may be used with concentrated sulfuric acid to initiate the above igniters and the following acid activated igniters: Sugar-Chlorate (0201), Fire Fudge (0202), and Match Head (0205). This delay may be used with Glycerin—Potassium Permanganate (0206).

(3) This device will produce fairly accurate delay times between one and eleven hours.

Caution: **The ticking sound of the clock may reveal the presence of the device.**

b. *Material and Equipment.*

Alarm clock, manually wound (without bell, if possible).
Bottle.
String.
Initiator liquid.
Cloth or absorbent paper.

c. *Preparation.*
  (1) Remove the bell or striker from the clock.
  (2) Fully wind time and alarm springs.
  (3) Set desired time on alarm.
  (4) Tie the string to the alarm key so that it will be pulled when the alarm mechanism is tripped. If necessary, tie a twig or stick to the alarm key to obtain a longer level.

d. *Application.*
  (1) Tie the string to the alarm key or stick. Set the clock in place and anchor it if necessary. Muffle the clock with rags, making sure that

the rags do not interfere with the reeling action of the alarm mechanism. Tie the free end of the string to the bottle of activating liquid. The bottle should be tilted in the direction of the fall by a pencil or twig. When this device is placed on a smooth surface, the clock should be taped, tied, or weighted down to prevent it from sliding when the tension in the string is taken up by the revolving key.

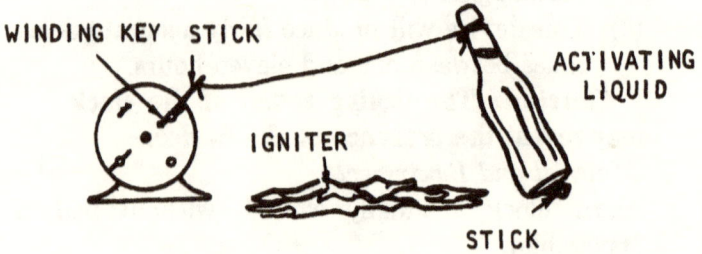

(2) Adjust the spacing so that the string is taut. Place a cloth or an absorbent paper towel where the contents of the bottle will be spilled. Place a quantity of igniter mixture on the cloth or paper towel. Partially overlap the igniter mixture with a flammable material so as to assist combustion.

CHAPTER 6
SPONTANEOUS COMBUSTION

0501. SPONTANEOUS COMBUSTION

a. Description.
(1) Spontaneous combustion is the outbreak of fire in combustible material that occurs without application of direct flame or spark. A combustible material such as cotton waste, sawdust, or cotton batting is impregnated with a mixture of a vegetable oil and specific drying oils known as driers. This impregnated combustible material is placed in a container which provides confinement around the sides and bottom. Heat produced by the chemical action of the driers in the oil is transferred to the confined combustible material with resultant outbreak of fire. Prepared igniter materials such as Fire Fudge (0202) or initiator material such as Fuse Cord (0101) can be used with the impregnated combustible material to increase reliability and decrease ignition delay time.

(2) The chemical reaction that supplies heat in the spontaneous combustion device becomes faster as the surrounding temperature rises. Conversely, as the temperature drops ignition delay time increases. In addition, ignition delay time varies somewhat with type of

vegetable oil, type of drier, type of combustible material, confinement, density of the oil impregnated combustible material, and ventilation. Devices planned for use should be tried in advance to establish delay time.

(3) These devices operate with a natural delay caused by the chemical reaction time of the drying process in the oil. The user places the device and is away from the scene when the fire starts. Spontaneous combustion devices have the added advantage of using items seen daily around shop, plant, or office. Containers for confining the impregnated combustible material can be small waste paper baskets, packing boxes, ice cream containers, paper bags and other items common to a particular operation. Combustible materials such as cotton waste, cotton batting, or sawdust are also common in many manufacturing plants. For these reasons, spontaneous combustion devices are useful and clever sabotage items.

(4) It is recommended that these devices be covertly used to ignite readily flammable material such as rags, dry paper, dry hay, wooden and cardboard boxes, wooden structures, and other similar targets.

b. *Material and Equipment.*

Ingredient	Used for	Common source
Vegetable Oils		
Boiled linseed oil	Paint manufacture	Hardware stores
Raw linseed oil	Paint manufacture	Hardware stores
Safflower oil	Food	Drug and food stores
Tung oil (China wood).	Paint manufacture	Paint manufacturers

Ingredient	Used for	Common source
Driers		
Cobalt (6%)	Paint manufacture	Paint manufacturers
Lead (24%)	Paint manufacture	Paint manufacturers
Manganese (can be substituted for cobalt).		
Lead oxide (can be substituted for lead).		
Combustible Materials		
Cotton waste	Machine shops, maintenance shops.	By-product of textile manufacture.
Cotton batting	Furniture manufacture.	Felt & textile manufacturers.
Sawdust	Water-oil-grease absorbent.	By-product of food working.
Kapok	Life jackets, furniture padding, bedding.	Furniture manufacturers, food products manufacturers.
Miscellaneous Items		
Cardboard or paper container.	General	Commonly available
Stick approximately 1½ inches in diameter.	General	Commonly available
Sharp knife	General	Commonly available
One pint wide-mouth jar.	General	Commonly available
Teaspoon	General	Commonly available
Fire Fudge Igniter (optional).	Igniter	See paragraph 0202
Fuse Cord (optional).	Initiator	See paragraph 0101

Proportions of Mixture

System	Vegetable oil	Cobalt drier (teaspoon)	Lead drier (teaspoon)	Combustible material (tightly packed)
1	Boiled linseed oil, 1/8 pint.	1/2	2	Cotton waste, 1 pint.
2	Boiled linseed oil, 1/8 pint.	1/2	2	Cotton batting, 3 pints.
3	Boiled linseed oil, 1/8 pint.	1/2	2	Sawdust, 1 pint
4	Boiled linseed oil, 1/8 pint.	1/2	2	Kapok, 1 pint
5	Raw linseed oil, 1/8 pint.	1	4	Kapok, 1 pint
6	Safflower oil, 1/8 pint.	1/2	2	Cotton waste, 1 pint.
7	Safflower oil, 1/8 pint.	1/2	2	Cotton batting, 3 pints.
8	Safflower oil, 1/8 pint.	1/2	2	Sawdust, 1 pint
9	Safflower oil, 1/8 pint.	1/2	2	Kapok, 1 pint
10	Tung oil, 1/8 pint.	1/2	2	Cotton waste, 1 pint.
11	Tung oil, 1/8 pint.	1/2	2	Cotton batting, 3 pints.
12	Tung oil, 1/8 pint.	1/2	2	Sawdust, 1 pint
13	Tung oil, 1/8 pint.	1/2	2	Kapok, 1 pint

Note. The above quantities for each system are approximately correct for use in a 1 gallon confinement container. The impregnated combustible material should fill the container to approximately 1/3 to 1/2 the volume for best results. Different size containers can be used with properly adjusted quantities of impregnated combustible material. At approximately 70° F., delay time to ignition is roughly 1 to 2 hours. With Fire Fudge or Fuse Cord added to the impregnated combustible material, delay time is reduced to roughly 1/2 to 1 hour. The exception to this is System 8 where delay time to ignition is about 2 to 3 hours. With Fire Fudge or Fuse Cord added, delay time is shortened to 1 to 2 hours.

c. *Preparation.*
  (1) *General instructions.*
    (a) Measure the combustible material by tightly packing it up to the top of the one pint measuring jar. The material should puff out of the measuring jar when firm hand pressure is removed.
    (b) Transfer the combustible material from the measuring jar to the container in which it is to be confined.
    (c) Pour the vegetable oil into the one pint measuring jar to one-third jar volume.
    (d) Using a teaspoon, add the specified quantity of Cobalt Drier to the vegetable oil in the one pint measuring jar. Wipe the spoon dry and add the specified quantity of Lead Drier to the Vegetable Oil—Cobalt Drier mixture.
    (e) Thoroughly mix the combination of vegetable oil and driers by stirring with the teaspoon for approximately one minute.

  Note. Vegetable oil and drier can be mixed and stored in an air-tight container for one week before use. Longer storage is not recommended.

    (f) Pour the oil mixture from the one pint measuring jar over the combustible material in the container. Saturate the combustible material by kneading, pulling and balling with the hands. This can be accomplished either inside or outside of the container.

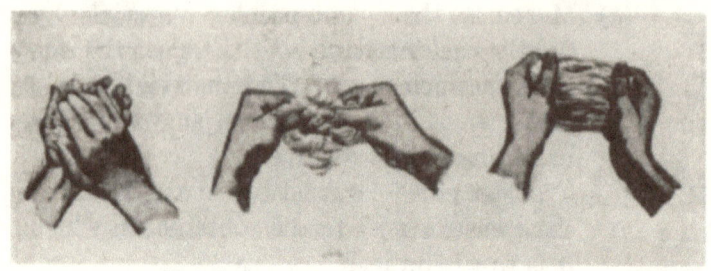

(g) Remove saturated combustible material from the container.

(h) Cut a hole with a knife, one to two inches in diameter, in the bottom center of the container.

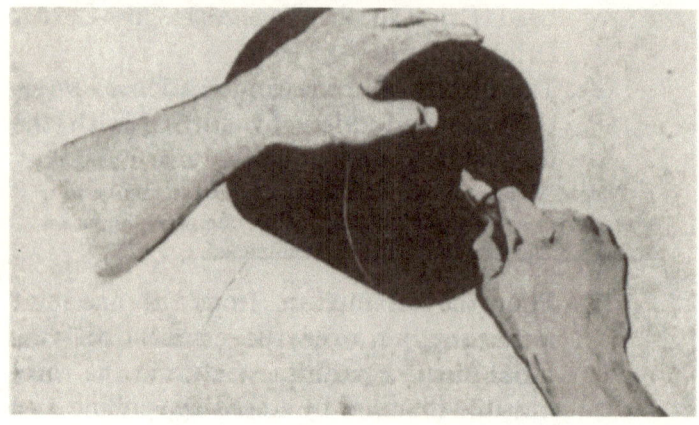

(i) Place the container on a flat surface, hold the 1½-inch diameter stick vertically over the hole in the bottom of the container and pack the saturated combustible material around the stick compressing it so that it fills ⅓ to ½ of the container volume after hand pressure is removed.

(j) Remove the stick. This leaves a ventilation hole through the center of the combustible material. The spontaneous combustion device is now ready for use unless the following optional step is taken.

(k) This step is optional. *Either* take a piece of Fire Fudge (0202) about the size of a walnut and crush it into pieces about the size of peas. Sprinkle the pieces of crushed Fire Fudge on top of the combustible material. *Or* cut a piece of Fuse Cord (0101) to a length of about four inches. Since safety fuse burns inside the wrapping, it is sliced in half to expose the black powder. (Lacquer coated fuse (nonsafety type) burns completely and may be used without slicing.) Insert one or more pieces of fuse vertically in the combustible materiel near the center vent hole, leaving about one inch extending out of the top surface of the combustible material.

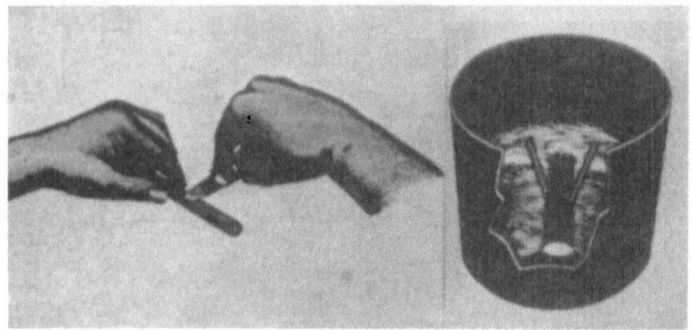

(2) *Preparation of improvised driers.* If the commercial driers (cobalt and lead) specified under *Material and Equipment* above are not available, the following improvised driers can be made using either flashlight batteries or powdered lead oxide (Pb_3O_4). These improvised driers are used in the same manner as the commercial driers.

  (a) *Manganese drier.*

    1. Break open three flashlight batteries (size

D) and collect the pasty material surrounding the central carbon rod.
2. Put this material in a one-pint wide-mouth jar and fill jar with water.
3. Slowly stir contents of jar for approximately two minutes and allow contents to settle. The contents will usually settle in one-half hour.
4. Pour off water standing on top of settled contents.
5. Remove wet contents from jar, spread it on a paper towel and allow to dry.
6. Dry the jar.
7. Pour raw linseed oil into the one-pint measuring jar to one-third jar volume.
8. Combine the measured quantity of raw linseed oil and the dried battery contents from *5* above in a pot and boil for one-half hour.
9. Shut off heat, remove pot from the heat source, and allow the mixture to cool to room temperature.
10. Separate the liquid from the solid material settled on the bottom by carefully pouring the liquid into a storage bottle. Discard the solid material. The liquid is the drier.
11. The manganese drier is ready for use.
12. If manganese dioxide powder is available, flashlight batteries need not be used. Place one heaping teaspoonful of manganese dioxide powder into the raw linseed oil and boil the mixture in a pot for one-half hour. Then follow *9*, *10*, and *11* above.

(b) *Lead oxide drier*.
1. Pour raw linseed oil into the one-pint measuring jar to one-third jar volume.
2. Combine the measured quantity of raw linseed oil and two heaping teaspoonfuls of lead oxide in a pot and boil gently for one-half hour. The mixture must be stirred constantly to avoid foaming over.
3. Shut off heat, remove pot from the heat source, and allow the mixture to cool to room temperature.
4. Pour the liquid into a storage bottle and cap the bottle.
5. The lead oxide drier is ready for use.

d. *Application*.
(1) The spontaneous combustion device is placed at the target on a flat surface with one edge propped up to allow ventilation through the impregnated combustible material.

Since flames normally shoot up from the open top of the container, combustible target material should be positioned from three to five inches directly over the top of the device for satisfactory ignition of the target. *DO NOT COVER OPEN TOP OF CONTAINER.*

(2) Temperature of the environment in which these devices are used affect, the ignition time these devices are used affects the ignition time. The following table gives approximate time to ignition at different temperatures. However, it is recommended that ignition time be determined by advance trial.

Temperature (° F.)	Time to ignition (hours)
60–70	1–2
40–60	2–4
30–40	4–10

(3) Spontaneous combustion devices can also be improvised by stuffing impregnated combustible material into a pocket of any one of the following garments: coat, laboratory jacket, pants, or similar items. The combustible material stuffed in the pocket should be below the top of the pocket and should not be packed too tight.

INDEX

	Paragraph	Page
Alarm clock delay	0412	128
Alcohol:		
Incendiary	0308	90
Lye thickeners for gasoline	0303.1 to 0303.3	57 to 63
Soap thickener for gasoline	0303.4	64
Aluminum powder igniter mixes	0204, 0207	30, 36
Animal blood thickener for gasoline	0303.8	74
Balancing stick delay	0410	121
Balsam-lye thickener for gasoline	0303.3	63
Barium peroxide—magnesium powder igniter.	0210	45
Blasting time fuse	0101	23
Brick, incendiary	0309	91
Candle delay	0406	109
Chemical hazards	0303	57
Chemical materials. (*See* specific materials, thickeners.)		
Cigarette delay	0401	94
Cobalt drier	0501	131
Combustible materials	0501	131
Corrosive action delay	0409	117
Definitions	0002	4
Delay mechanisms. (*See also* specific item.)	0306, 0401 to 0412	82, 94 to 128
Diaphragm delays	0403, 0404	101, 105
Dissolving action delay	0409	117
Egg thickener for gasoline	0303.5	66
Fire:		
Bottles	0305, 0306	78, 82
Fudge igniter	0202	25
Hazards	0003	5

	Paragraph	Page
Flammable liquids	0308	90
Fuse:		
Cords	0101, 0102	8, 14
Lighter, M2	0101	8
Lighter, M60	0101	8
Gasoline incendiaries	0302, 0303, 0308	53, 57, 90
Gelatin capsule delay	0402	98
Gelled gasoline incendiaries	0302, 0303	53, 57
Glycerin delays	0405, 0407	107, 111
Glycerin—potassium permanganate igniter.	0206	34
Igniters. (*See also* specific item.)	0201 to 0211	23 to 47
Incendiary:		
Brick	0309	91
Materials. (*See also* specific item.)	0301 to 0309	50 to 91
Systems	0001	3
Initiators. (*See also* specific item.)	0101 to 0104	8 to 21
Latex thickener for gasoline	0303.6	69
Lead drier	0501	131
Linseed oil	0501	131
Lye thickeners for gasoline	0303.1 to 0303.3	57 to 63
Magnesium powder igniter mixes	0208, 0210	39, 45
Match head igniter	0205	32
Napalm incendiary	0301	50
Oil of vitriol initiator	0103	16
Overflow delay	0407	111
Paper diaphragm delays	0404, 0405	105, 107
Paraffin-sawdust incendiary	0304	76
Potassium:		
Chlorate-sugar igniter	0201	23
Permanganate crystals delay	0405	107
Permanganate—glycerin igniter	0206	34
Rubber band delay	0411	126
Rubber diaphragm delay	0403	101

	Paragraph	Page
Safety Fuse, M700	0101	8
Safflower oil	0501	131
Silver nitrate—magnesium powder igniter.	0208	39
Soap-alcohol thickener for gasoline	0303.4	64
Sodium:		
Chlorate-sugar igniter	0201	23
Peroxide—aluminum powder igniter.	0204	30
Peroxide—sugar igniter	0203	28
Spontaneous combustion	0501	131
String fuse	0102	14
Subigniter for thermite	0211	47
Sugar igniter mixes	0201, 0203	23, 28
Sulfur pellets—aluminum powder igniter.	0207	36
Sulfuric acid delays	0402 to 0404	98 to 105
Sulfuric acid initiator	0103	16
Thermite igniter	0211	47
Thermite incendiary	0307	85
Thickeners for gasoline. (*See also* specific material.)	0302, 0303	53, 57
Tipping delays	0408 to 0410	113 to 121
Tools and techniques. (*See also* specific device.)	0003	5
Tung oil	0501	131
Water delays	0402, 0407 to 0409	98, 111, 117
Water initiator	0104	21
Wax thickener for gasoline	0303.7	71
White phosphorus igniter	0209	41

TAGO 7189-C

By Order of the Secretary of the Army:

HAROLD K. JOHNSON,
General, United States Army,
Chief of Staff.

Official:
J. C. LAMBERT,
Major General, United States Army,
The Adjutant General.

☆ U. S. GOVERNMENT PRINTING OFFICE : 1967 O - 300-528 (7016C)

IMPROVISED INCENDIARIES
General

Good incendiaries can be improvised more easily than explosives and the materials are more easily obtained. On a pound for pound basis, incendiaries can do more damage than explosives against many type targets if properly used. There is a time lag, however, between the start of a fire and the destruction of the target. During this period the fire may be discovered and controlled or put out. An explosive once detonated has done its work.

Incendiaries are cheap and little training is needed for their preparation and use. Used in very carefully excuted operations, the act of sabotage may be concealed in the ashes of an "accidental" fire.

Fires may be started quickly and have reasonable chance of success if the following few simple principles are observed:

1. See that there is plenty of air and fuel to feed the fire.
2. Use an incendiary that supplies a prolonged and persistent heat.
3. Start the fire low in the target structure and let it spread naturally upwards.
4. Use reflecting surfaces, such as corners, boxes, shelves, to concentrate the heat.
5. Use drafts to spread the fire rapidly — near stairways, elevator shafts.
6. Protect the fire from discovery during the first few minutes by good concealment and timing.

In preparing improvised incendiaries observe basic rules of safety. Chemicals that must be powered should be ground separately with clean tools and then mixed in the indicated proportions. Chemicals or mixtures should be kept tightly sealed in jars or cans to protect them from moisture. Damp materials will work poorly if at all.

Sulfuric acid, which is useful for chemical delays and to ignite incendiaries or explosive detonators, can be obtained by concentrating battery acid. This can be done by boiling off the water in the battery acid in a glass or porcelained pan until dense white fumes begin to appear. This operation should be done out of doors and the resulting concentrated acid should be handled carefully.

The paragraphs which follow will describe the preparation of several igniter (or "first fire") incendiary mixes, some basic incendiary mixes, and a thermate metal-destroying incendiary.

The subject of incendiaries has been treated much more exhaustively in other publications. The intent of this handbook is to provide only a few techniques.

Potassium Chlorate and Sugar Igniter

Chlorate-sugar is one of the best of the first fire or igniter mixes. It burns very rapidly, with a yellow-white flame, and generates sufficient heat to ignite all homemade incendiaries mentioned in this handbook.

MATERIALS: Potassium chlorate (preferred) or sodium chlorate, sugar.

PREPARATION:

1. Grind the chlorate separately in a clean, non-sparking (glass or wooden) bowl with a wooden pestle. the resulting granules should approximate those of ordinary table sugar.

2. Mix equal volumes of the granulated chlorate and sugar by placing both on a large sheet of paper and then lifting the corners alternately.

CAUTION: This mixture is extremely spark sensitive and must be handled accordingly.

3. Wrap 4 to 6 tablespoonfuls of the mixture in thin paper so as to form a tight packet. Keep the mixture as dry as possible. If it is to be stored in a damp area before using, the packet may be coated with paraffin wax.

Chlorate-sugar is easily ignited by the flame of a match, the spit of a percussion cap or time fuse, with concentrated sulfuric acid.

If ignited when under confinement it will explode like gunpowder. If it is contained in a waxed packet, therefore, the latter should be punched through in several places before it is used with a basic incendiary and ignited.

Flake Aluminum-Sulfur Igniter

This simple igniter burns extremely hot and will ignite even the metal-destroying thermate, described later on. The mixture itself can be lit by chlorate-sugar.

MATERIALS: Flake aluminum, finely powdered sulfur.

PREPARATION:

1. Mix 4 parts by volume of finely powdered sulfur with 1 part of aluminum powder.

To use, place several spoonfuls of the mixture on the material to be lit and add a spoonful of chlorate-sugar on top. Be sure the safety (time) fuse or other spark-producing delay system is placed so it will act upon the chlorate-sugar mixture first.

Homemade Black Powder Igniter

Black powder may be used for igniting napalm, flammable solvents in open containers, paper, loose rags, straw, excelsior and other tinder type materials. If it is not available already mixed, it can be prepared as follows:

MATERIALS: Potassium (or sodium) nitrate, powdered charcoal. powdered sulfur, powde)

PREPARATION:

1. Into a clean, dry jar or can put 7 spoonfuls of potassium or sodium nitrate, 2 spoonfuls of powdered charcoal, and 1 spoonful of powdered sulfur. The ingredients must be at least as fine as granulated sugar. If they must be ground, GRIND EACH SEPATATELY. Never grind the mixed ingredients — they may ignite or explode.

2. Cap the can or jar tightly and shake and tumble it until the ingredients are completely mixed.

The mixture will be effective for months if kept tightly sealed and dry. Sodium nitrate in particular has a tendency to absorb moisture.

To use the gunpowder, pile 2 or 3 spoonfuls on top of any solid incendiary material which is to be ignited. For igniting liquids in open containers, wrap 2 or 3 spoonfuls in a piece of paper and suspend it just above the liquid.

Gunpowder is best ignited by safety fuse. It burns very quickly and with a great deal of heat, so allow sufficient time delay for safe withdrawal from the vicinity.

Match Head Igniter

A good ignition material for incendiaries can be obtained from the heads of safety matches, which are available almost any place. The composition must be removed from the heads of many of them to get a sufficient quantity of igniter material. It will ignite napalm, wax and sawdust, paper, and other flammables.

MATERIALS: Safety matches.

PREPARATION:

Remove the match head composition by scraping with a knife or crushing with pliers. Collect several spoonfuls of it and store in a moisture-tight container.

Put at least 2 spoonfuls on the material to be ignited. To ignite liquids, such as solvents or napalm, wrap several spoonfuls in a piece of paper and hang this just over the fluid, or place nearby. If fluids dampen the mixture it may not ignite.

Ignition can be by time fuse, fircracker fuse, a spark, or concentrated sulfuric acid.

Time Fuse Fire Starter

Several igniters or first fire mixes can be set off by a spark from time fuse. Others require a stronger flame. Time fuse, plus matches, can be combined to improve this more intense initial flame.

MATERIALS: Time (safety) fuse, safety matches, string or tape.
PREPARATION:

1. About ¼ inch from the end of a piece of time fuse cut a notch with a sharp knife so that the powder train is exposed.

2. Around the fuse at this point tape or tie several matches so that their heads are in contact with each other and at least one match head is directly over the notch. See Figure 59.

When the fuse burns down, a spark from the notch ignites the one match head, which flares and ignites the others. this fire starter can be inserted into an igniter mix or used alone to light crumped paper or excelsior. Another application, nonelectric firing of the 3.5" rocket, is described earlier.

Homemade Napalm

Napalm is the best incendiary to use against most flammable targets. It will readily ignite paper, straw, flammable solvents, or wooden structures.

MATERIALS: Gasoline or fuel oil, nondetergent soap (bar, flakes, or powder).

PREPARATION:

1. Use about equal parts of soap and oil. If bar soap is used, slice it into small chips. If both gasoline and fuel oil are available, use both in equal parts.

2. Heat the fuel in an open container, preferably one with a handle, out of doors. Try to avoid creating sparks or having a high open flame, but if the fuel should catch on fire extinguish it by placing a board or piece of tin over the container.

3. Gasoline, in particular, will begin to bubble very quickly. When it does, remove from the fire and gradually add the soap, stirring continuously, until the soap is completely dissolved and a thin pasty liquid results. If necessary return the mixture to the fire, but as a safety measure it is best not to stir while the container is on the fire.

4. When the desired consistency is reached allow the mixture to cool.

5. Napalm also can be mixed by a cold method, although it may take hours to thicken. This should be done by alternately adding very small amounts of soap chips or powder and gasoline or fuel oil and stirring until the mixture reaches a thin jelly-like consistency. It is best to start with

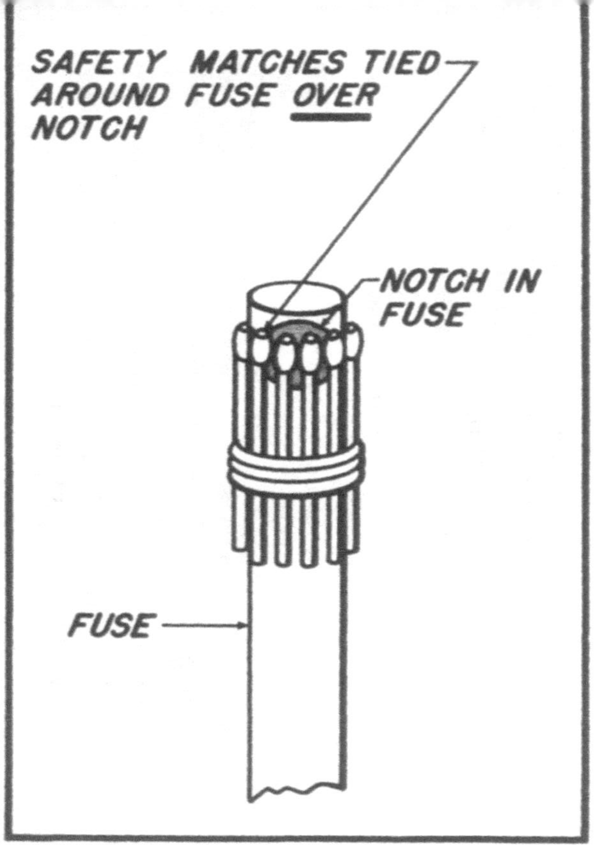

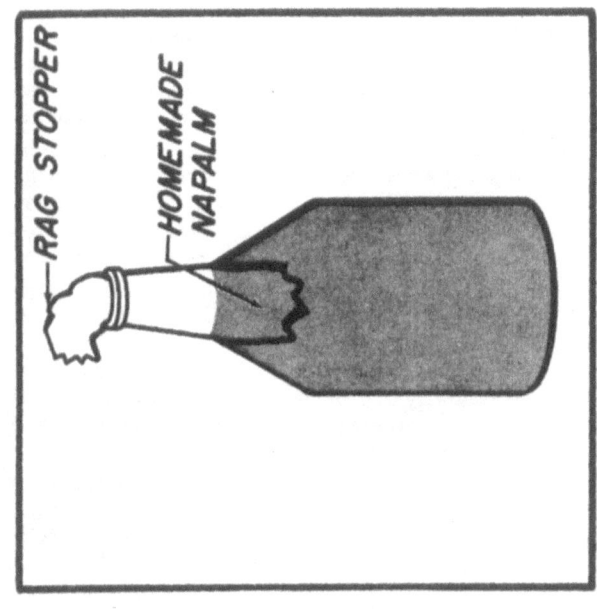

Fig. 59 — Time Fuse Fire Starter

Fig. 60 — Molotov Cocktail

about a cupful of soap, add part of a cup of solvent and stir that until smooth before gradually adding the remaining ingredients. Continuous stirring is not required. In fact, it is advisable just to let the mixture and the mixer rest from time to time and give the soap a chance to dissolve.

Napalm will keep well if stored in a tightly sealed container. It can be ignited with a match or any of the first fire mixtures described previously. The ignition packet should be placed adjacent to or just over the napalm, otherwise the petroleum may soak it and prevent its burning.

When napalm is used on easily ignitible materials (such as loosely piled paper, rags, or hay) it should be spread out so it will start a large area burning at once. Tightly baled paper or rags should be loosened first, because they do not burn well. If used directly against wooden structures or other large articles which are difficult to ignite, the napalm should be concentrated in sufficient quantity to provide a hot, long lasting blaze. If about a half dozen pieces of charcoal are put into and around the napalm the heat output is considerably increased.

Napalm makes an excellent "Molotov cocktail." Just fill any glass bottle with a small neck with the napalm and cram a twisted strip of cloth into the mouth of the bottle as a stopper. See Figure 60. When ready to use, pull about 4 to 6 inches of the rag stopper out of the bottle. Light the napalm-soaked rag with a match and, when the rag is burning well, throw the bottle at the target. When the bottle breaks napalm spashes over the target and is ignited by the burning rag.

Paraffin-Sawdust Incendiary

Paraffin-sawdust is almost as effective as napalm against combustible targets, but it is slower in starting. It is solid when cool and thus is more easily carried and used than liquid napalm. In addition, it can be stored indefinitely without special care.

MATERIALS: Dry sawdust, paraffin, beeswax, or candle wax.

PREPARATION:

1. Melt the wax, remove the container from the fire and stir in a roughly equal amount of sawdust.

2. Continue to stir the cooling mixture until it becomes almost solid, then remove from the container and let it cool and solidify further.

Lumps of the mixture the size of a fist are easiest to manage. The chunks of incendiary may be carried to the target in a paper bag or other wrapper. Any igniter that will set fire to the paper wrapper will ignite the wax and sawdust.

A similar incendiary can be made by dipping sheets of newspaper into melted wax and allowing them to cool. These papers may then be crumpled up and used in the same manner as the paraffin-sawdust, although they will not burn as hot and persistently.

Sawdust, Moth Flakes, and Oil Incendiary

This incendiary is very good for use against all kinds of wooden structures, including heavy beams and timbers. It also works well on paper, rags, straw, excelsior, and other tinder type materials. It will start fires in open containers of flammable liquids, piles of coal, coke, or lumber, and on baled rags and paper. It is not effective against metal.

MATERIALS: Dry sawdust, moth flakes (naphthalene), fuel oil (kerosene or diesel oil).

PREPARATION:

1. Place equal parts of sawdust, moth flakes, and oil into a container and stir until the mixture is the consistency of mush.

2. Store it in any container that will retain the oil fumes.

An easy, effective way to use this mixture is to put about a quart of it in a paper bag and place the bag on the target material. The bag can be lit with a match and the mixture will ignite quite readily. It burns as well as napalm. If a longer delay time is required, use one of the igniter mixes described earlier along with time fuse or other delay device. The time fuse alone, however, will not ignite the incendiary mix.

Where very large wood beams are to be burned, an additional amount of the incendiary will be required. Two or three quarts is enough to destroy almost any target against which the technique would be effective.

For the greatest effect on wooden structures, the mixture should be in a pile, never spread out in a thin layer. It should be placed beneath the target material, if possible, so the flames will spread upward. In a packing box or room, a corner is a good place to start the fire.

Thermate Incendiary

Thermate is similar to commercial thermit, used in welding, except that it also contains an oxidizer, making it easier to ignite. Thermate will readily burn paper, rags, excelsior, straw, and other tinder type materials. However, its main use in sabotage operations is against motors, gears, lathes, or other metal targets — to weld moving parts together, warp precision machined surfaces, and so on. Since it burns with a brief, almost explosive action, it is not recommended for burning wooden structures or other materials where persistent heat is required.

A good source of ready-made thermate is the U.S. military AN M-14 Incendiary Grenade. To remove the thermate, first pry out the fuse assembly with crimpers or other nonsparking implement. See Figure 61. The reddish-brown caked substance on top of the contents of the grenade is a first fire mixture and it is spark sensitive. This should be broken up and the grayish powder beneath, which is the thermate, can be poured out.

Thermate also can be made from aluminum or magnesium powder and a chemical oxidizing agent, as described below:

MATERIALS: Aluminum filings, powder or flakes, or magnesium filings or powder, plus any one of the following chemicals: potassium nitrate, sodium nitrate, barium nitrate, potassium dichromate, sodium dichromate, or potassium permanganate. Although aluminum and magnesium are equally effective, thermate made from magnesium is easier to ignite. Flake aluminum, which is the extremely fine variety used in paints, is excellent. In any case, both the metal and chemical ingredients should be no coarser than granulated sugar.

PREPARATION:

1. Fill a quart size (or larger) container about 2/3 full of equal parts of the metal powder and the oxidizing agent.

2. Cover with a tight lid, then roll and tumble the container until the contents are completely mixed.

3. If flake aluminum is the metal used, fill the container ½ full of the aluminum then add oxidizing agent until the container is ¾ full. Mix as described above.

Thermate in a sealed container can be stored for months. To use, put 1 or 2 pounds of the mixture in a paper bag and place it on the target in such a way that when it burns the red hot molten material will run down and attack the vital parts.

Chlorate-sugar and aluminum-sulfur igniters are best for setting off thermate, particularly if the thermate contains aluminum powder, which is more difficult to ignite.

Thermate also is used in the improvised dust initiator and the external POL charges described later.

Fig. 61 — Defusing Thermate Grenade

www.ingramcontent.com/pod-product-compliance
Lightning Source LLC
Chambersburg PA
CBHW030741180526
45163CB00003B/877